ELECTRONICS IN EXPERIMENTAL PHYSICS

ELEKTRONIKA V FIZICHESKOM EKSPERIMENTE

ЭЛЕКТРОНИКА В ФИЗИЧЕСКОМ ЭКСПЕРИМЕНТЕ

The Lebedev Physics Institute Series

Editor: Academician D. V. Skobel'tsyn

Director, P. N. Lebedev Physics Institute, Academy of Sciences of the USSR

Proceedings (Trudy) of the P. N. Lebedev Physics Institute

Volume 42

ELECTRONICS IN EXPERIMENTAL PHYSICS

Edited by

Academician D. V. Skobel'tsyn

Director, P. N. Lebedev Physics Institute
Academy of Sciences of the USSR, Moscow

Translated from Russian

SPRINGER SCIENCE+BUSINESS MEDIA, LLC 1970

The Russian text was published by Nauka Press in Moscow in 1968 for the
Academy of Sciences of the USSR as Volume 42 of the Proceedings (Trudy)
of the P.N. Lebedev Physics Institute. This translation is published under
an agreement with Mezhdunarodnaya Kniga, the Soviet book export agency.

Library of Congress Catalog Card Number 69-12524

ISBN 978-1-4757-5111-6   ISBN 978-1-4757-5109-3 (eBook)
DOI 10.1007/978-1-4757-5109-3

© 1970 Springer Science+Business Media New York
Originally published by Plenum Publishing Corporation, New York in 1970
Softcover reprint of the hardcover 1st edition 1970

All rights reserved

No part of this publication may be reproduced in any
form without written permission from the publisher

CONTENTS

THE LABORATORY MEASUREMENT
AND REGISTRATION CENTER

I. V. Shtranikh, V. N. Bochkarev, A. N. Volkov, A. M. Klabukov, and V. V. Puzanov

Introduction

The notion advanced at the Fifth Scientific and Technical Conference on Nuclear Radio Electronics in 1961 of creating a centralized facility for the measurement and registration of any and all spectrometric information in a special laboratory [1] has "come of age," having begun to materialize into a manifest and feasible reality. Furthermore, both in the Soviet Union and in other countries we already find examples of such facilities that have been completed or are in the process of being built, in most cases on a computerized basis [2-11].

The situation just described is a logical one and stems from the fact that the effective and often the only solution to many problems in nuclear physics cannot be realized other than by methods of multivariate (multiparameter) analysis [8]. The processing of incoming spectral data must be accomplished on computers.

In large-scale laboratories equipped with several accelerator devices and staffed by many physics user groups it would be highly impractical to meet simultaneously the demands of all user groups for the processors needed. Such equipment is still bulky and complex. Assigned to a special group, it would be used inefficiently on the average.

In our opinion, the sensible approach to this situation is to build a unified spectrometric measurement facility [1] having a large handling capacity and serving all user groups of the laboratory. Such a facility might be conceived in the future as a complex of several multidimensional recording instruments connected directly into a special-purpose computer.

This kind of centralization has the advantage of: 1) reducing the volume of equipment used (particularly with regard to data output); 2) amplifying the possibilities of setting up more complex physical problems in which a very large number of channels is required; 3) increasing the equipment load capabilities; 4) simplifying equipment servicing operations and enhancing its operational reliability through the unification of instrumentation.

The facility can be built on the basis of a typical medium-capacity (eight to ten thousand word) computer. Until 1964, however, these possibilities were practically out of the question. Nevertheless, some individual computer modules were made available. Therefore, systems having a capacity of more than ten thousand channels can be designed with a minimum of effort on the basis of standard computer equipment. Best suited to this purpose are large ferrite storage devices and magnetic drums.

Very simple systems utilizing semipermanent storage of the magnetic tape variety are less preferable and should be used in the facility as readout devices, for the accumulation of statistics, and for the subsequent rewriting and sorting of all multivariate data. Magnetic tape does not permit real-time observation of the overall volume of recorded data, hence its use in research projects is limited.

We have used storage devices of the magnetic drum type with 88 tracks. This type of device is readily suited to the construction of systems with ten thousand channels or more and a channel capacity of $2^{14} - 1$ readings. The advantages of a drum system include the capability of continuous observation of all data, flexibility in the allocation of its capacity, and fast access to its individual groups of channels (tracks). The drawback of a conventional system of this type is its slow response time. This drawback is eliminated with the incorporation of special buffer memory devices of the "flattening" (derandomizing) type with a code rewrite capability.

A series of papers [12-17] has been devoted to the detailed description of one of two almost identical measurement and registration center (MRC) equipment systems installed in two laboratories. In contrast with the earlier project described in [1], the system in this case underwent major modifications in the course of its full realization.

1. The Proposed MRC System

A block diagram of the center is shown in Fig. 1. In the ensuing paragraphs we consider only those functions of the MRC which depart from the conventional. The incoming data records of events are delivered along cables in analog form from the instrumentation groups to the input patchboard 1 of the center. From there the event signals are transmitted through amplification and transmission networks and are coded (in amplitude, time, angle, etc. [12, 13, 18]) by the encoders 2, where they are stored for a certain period of time in code form. Time encoders are being developed with a capability of storing several code numbers at once (first buffer memory [13]).

An end-of-coding signal is transmitted via buses into the systems of the second buffer memory 3, which is of the "flattening" (derandomizing) type with rewriting of the numbers associated with given encoders. If in this memory there is a group of elements available to receive a code, the command SEND CODE is transmitted via another bus into block 2, encoder 2 is cleared, and the number is admitted to block 3.

Two memory systems of this type have been developed [14], one with direct selection of the available memory elements and direct readout of the loaded elements, and one with cyclic scan-rewrite interrogation of the element groups. Each of these devices is capable of handling five or more numbers. In the former system capacitor storage is used, in the latter system ferrite shift registers with a cycling frequency of about 90 kc are used for storage.

In the capacitor system, as opposed to [19], a simple combination of capacitances and common tunnel diodes is used. The charge-discharge current to the storage capacitor for each digital position controls the operation of a tunnel diode, and the latter initiates storage through a transistor. The shift register system functions analogously. The register output signals control the operation of the tunnel diodes.

The essential feature of these devices that sets them apart from earlier models [20-24] is their capability for backup and rewrite of the next code in the event it cannot be read at a given instant. This rewrite feature is inherent in the operation of the magnetic drum. All number code combinations (from 0 to 127) have a definite location on the semiperiphery of the drum 7. A conventional digital storage system is used [15]. Each track has two heads, a write head 8 and read head 9, which are 180° apart in phase (allowing for the phase lead of the read head). This arrangement cuts in half the main memory access time, which is about 1/50 sec (although

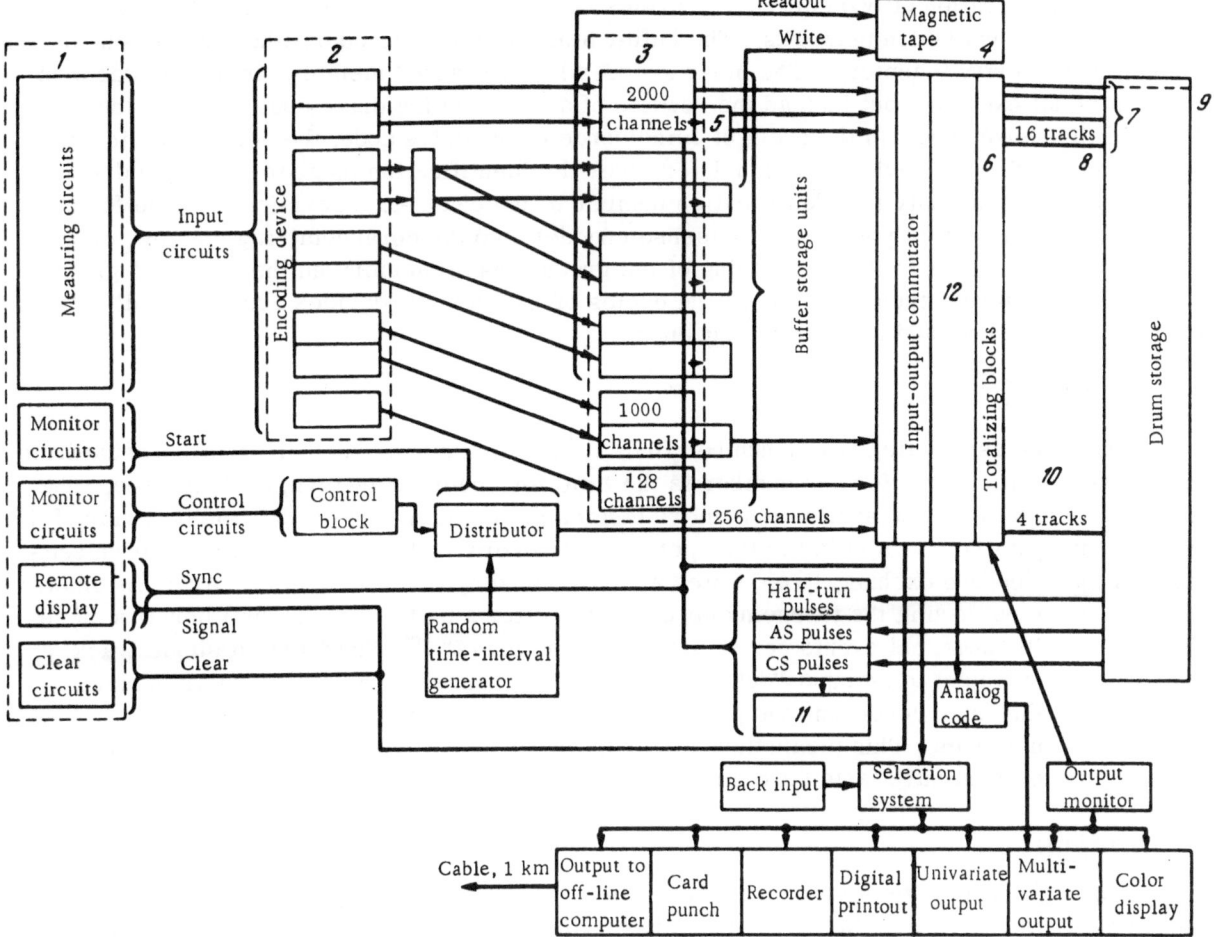

Fig. 1. Block diagram of the laboratory MRC. AS) Arithmetic sequence; CS) channel sequence.

it also cuts in half the capacity of the system), and in addition will permit the eventual elimination of the erase head, thus reducing the number of logic operations required in totalizing.

The number of identical code combinations is summed by the channel logic networks 6. Each channel, i.e., sector for the position of a given code combination, has a capacity of up to $2^{14}-1$ readings. For any specified angle of the drum semiperiphery, i.e., for a definite location of numbers from 0 to 127, a test is made for the presence of these numbers in the memory circuits 3 (in their code comparison circuits). A drum angle-of-rotation code read by the central station network 11 is delivered with phase lead to the aforementioned memory circuits by registration of the channel gate-pulse sequence.

During registration of one channel (about 160 μsec) 14 numbers can be scanned in each of the buffer storage units. If one of these combinations should coincide during that period, some of the memory circuits 3 send a command to the networks 6 to increment (or decrement) once this channel, which is immediately thereafter energized by the read head. At the same time the code is deleted from memory (3). If the code does not arrive, it is rewritten. Provision is made in the presence of several identical numbers (two or four) in a particular buffer unit for the immediate write-in of ±2 or ±4. It is possible to clear data to "0" according to the priority of the user groups. Consequently, this type of memory system is associative with respect to the main drum memory.

The registration of two correlating events is realized with two encoders having interconnected input transmission networks. The buffer storage of one encoder, like the one considered above, is the "drive" memory. The other encoder has a "sense" memory, in which code comparison is not executed, but instead the rewrite of a nonincoming drive code controls the rewrite of the sense code. In the event of coincidence of the drive code, the sense code is sent out to the distributing device 5, which determines the identification number (#) of the drum track and simultaneously the identification number of the channel. Provision is made for transfer of the leading digital position of the sense encoder into the derandomizing memory of the drive section. This permits partitioning of the gradations of memory on the axes into groups of 64 and 32 channels each. The operating cycles of the memory devices are "coupled" to the drum-stored channel and arithmetic sequences.

2. Response Time

Inasmuch as practically any number finds its location during one half-rotation of the drum, the average waiting time for any number is the time for one quarter-turn, or 1/100 sec. The incoming numbers fill the available memory elements, and, unlike [3, 5, 25], the behavior of the system as a whole is similar to the operation of general-subscription central-station sections of the telephone exchange type, which are described by queueing theory. Analysis shows [14] that a system having the foregoing specifications (for a uniform spectrum) permits more than 100 to 200 statistical events per second to be written (at 1% error) onto a single track, a rate that is equivalent to an input "dead time" of 100 to 50 μsec. For multivariate measurements under the same conditions the probability of a wanted number occurring in a given sector of the drum increases. This means that the input networks (if they are not to suffer a loss of response time) must have faster circuits with response times below 10 μsec.

3. Coding

For multivariate amplitude measurements a "position-weighting" encoder has been designed [18]; its tube-transistorized version has a differential nonlinearity of about 1% per channel for 128 channels. The coding time is less than 10 μsec. Since the entire memory system sums the codes, operating asynchronously with respect to the arrival of incoming information, it is possible to use any encoding devices with appropriate prestorage buffer sections. All that matters is that no more than 50 (or 100) statistical pulses per second be fed into one channel, bearing in mind also the limitation imposed above on the total number of pulses.

4. Data Output

Data output systems have been designed [16] using standard high-speed printout (20 channels/sec), standard card punch (10 channels/sec), recorders for individual channel output, oscilloscope display of any channel in digital or analog form, and an interface system for the direct transfer of data by cable to the institute digital computer at a rate of more than two thousand channels per second. In the data output system extensive use is made of plain and counting-type triggers. The latter type comprises a tunnel diode connected to a transistor, which controls the switching networks through a transformer.

A central-station system has also been developed for the color monitoring of all accumulated data. This system incorporates an ordinary mask plate color tube, on which each channel has allocated to it a single point, the color of which permits the number carried by the channel to be evaluated with 10% error. The overall topographical set of points (several defocused) affords something similar to a geographical map, the largest numbers in the channels being represented by red, the smallest by blue, etc. To accomplish this, a dynamic sequence of coded messages is sent from each track to its own associated simple five-position preliminary ferrite matrix (12 in Fig. 1).

Fig. 2. Overall view of the measurement and registration center.

All the matrices are serially scanned during the period of one channel. For this purpose the codes stored in them are counted by a series of blocking oscillators controlled by pulses from a delay line. Ultimately the common output consists of a train of parallel five-place pulse codes separated by an interval of 0.9 μsec. In order to facilitate formation of the pulses, during each half-turn of the drum writing is executed in only a fourth of the matrices, and readout takes place in groups of four each. The digital position at which readout is to begin is chosen at will for all channels simultaneously.

From the bus outputs the time sequence of pulses is transformed by the network 12 into analog form and is then fed to the circuit of a three-dimensional display oscilloscope, as well as to the color output network (after passing the decoding matrix), where it is transformed into the three main color components. Interlacing of the video scan makes it possible to have 256 horizontal lines, 64 assigned to each half-frame (half-turn of the drum).

To the right of the raster on the color tube screen (after 80 horizontal points) it is proposed that a control zone be included, its color varying from top to bottom. The zone is covered by a digital mask plate. The color of the zone is determined by the level of the interpolated voltage taken from the channel counter 11. The zone is used for a more precise determination (by comparison of the registered points with the color of the reference mask) of the numbers carried by the channels. The visual display of data can be duplicated at the disposal of the individual user groups.

5. Back Input of Data

A system has been designed [16] for the back input of data from punched cards through a standard reading device. The system permits the input of data with any coefficient, a feature that is important from the standpoint of background normalization, etc. The input data can be added to or subtracted from information carried earlier by the corresponding channels. The chosen coefficient must be expanded in powers with base two, and the cards are read as many times as there are powers in the coefficient. The system is readily adaptable to input from the tape units 4, which accumulate statistics for certain multivariate events, and it can write information from a large number of channels, in excess of the total system capacity.

Central station systems have also been designed for the monitoring and calculation of the "on-time" of the encoding devices. For this, four tracks are tapped on 256 channels with a channel capacity of $2^{28} - 1$ readings per channel (10).

Conclusion

Systems of the type described have been installed in two laboratories. The facility is set up so that its subsystems can be serviced and maintained by medium-skilled technicians, while the fabrication of necessary auxiliary equipment is handled within the capabilities of the institutes themselves. Thus considerable attention has been given over to ready accessibility to the various networks. For this purpose the equipment is housed in special multiaccess cabinets. Excessive compactness is avoided in the layout of the electronic components within the networks. The units are almost completely transistorized. An overall view of the center (MRC) is shown in Fig. 2.

The concept of centralized measurement and registration of spectra is also beginning to enjoy widespread utilization in other institutes throughout the Soviet Union and, given the proper approach, fully justifies its existence.

LITERATURE CITED

1. I. V. Shtranikh, V. N. Bochkarev, A. N. Volkov, and A. M. Klabukov, Proc. Fifth Sci.-Tech. Conf. Nuclear Radio Electronics, Vol. 2, Part 2, Gosatomizdat, Moscow (1963), p. 135.
2. G. I. Zabiyakin, V. N. Zamrii, and V. I. Semashko, Pribory i Tekh. Éksperim., No. 4, p. 139 (1964).
3. J. Lendg and A. Pearson, Internat. Symp. Nuclear Electronics, Paris (1963), p. 182.
4. I. V. Shtranikh, V. N. Bochkarev, A. N. Volkov, V. M. Geraseev, A. M. Klabukov, V. V. Puzanov, and A. M. Shimanskii [Shimansky], Internat. Symp. Nuclear Electronics, Paris (1963), p. 587.
5. G. Krüger and G. Dimmer, Internat. Symp. Nuclear Electronics, Paris (1963), p. 263.
6. B. E. Zhuravlev and G. I. Zabiyakin, Pribory i Tekh. Éksperim., No. 2, p. 81 (1966).
7. R. I. Spinrad, IEEE Trans., No. 3, p. 324 (1964).
8. B. G. Minaev, Yu. V. Stupin, and I. V. Shtranikh, FIAN Preprint A-149 (1965).
9. B. Souček, Rev. Sci. Instr., 36:750 (1965).
10. G. P. Zhukov, B. E. Zhuravlev, G. I. Zabiyakin, and V. N. Zamrii, Pribory i Tekh. Éksperim., No. 6, p. 34 (1964).
11. M. N. Ivanov, V. I. Kadashevich, I. A. Kondurov, S. N. Nikolaev, A. P. Nekhai, A. G. Nikanorov, and V. I. Petrova, Proc. Sixth Sci.-Tech. Conf. Nuclear Radio Electronics, Vol. 3, Gosatomizdat (1965), p. 45.
12. I. V. Shtranikh and A. N. Volkov, Proc. Fifth Sci.-Tech. Conf. Nuclear Radio Electronics, Vol. 2, Part 1, Gosatomizdat, Moscow (1963), p. 10.
13. V. V. Puzanov and I. V. Shtranikh, Trudy FIAN, 42:27 (1968). [This volume, p. 27.]
14. V. V. Puzanov, I. V. Shtranikh, and A. T. Matachun, Pribory i Tekh. Éksperim., No. 3, p. 82 (1966).
15. A. M. Klabukov and I. V. Shtranikh, Trudy FIAN, 42:53 (1968). [This volume, p. 53.]
16. A. M. Klabukov and I. V. Shtranikh, Trudy FIAN, 42:10 (1968). [This volume, p. 8.]
17. V. V. Puzanov and I. V. Shtranikh, Trudy FIAN, 42:43 (1968). [This volume, p. 43.]
18. A. N. Volkov, G. I. Zabiyakin, V. G. Tishin, and I. V. Shtranikh, Trudy FIAN, 42:33 (1968). [This volume, p. 33.]

19. Y. Amram, Nuclear Electronics, Vol. 2, IAEA, Vienna (1962), p. 73.
20. A. E. Voronkov, L. N. Korablev, I. D. Murin, and I. V. Shtranikh, The BMA-20 High-Speed Multichannel Analyzer, VINITI, Moscow (1957).
21. R. E. Bell, Canad. J. Phys., 34:563 (1956).
22. V. A. Vyazemskii, Author's abstract of Dissertation, LÉTI (1960).
23. G. P. Zhukov, G. I. Zabiyakin, V. D. Shibaev, and I. V. Shtranikh, OIYaI Preprint, No. 731 (1961).
24. T. K. Alexander and J. Lendg, CREL, 1036, Chalk River, Ontario (1961).
25. C. D. Goodmann, G. D. O'Kelley, and D. A. Bromley, Proc. Symp. Nucl. Instr., Harwell (1961).

DATA OUTPUT DEVICES IN THE SYSTEM OF THE MEASUREMENT AND REGISTRATION CENTER

A. M. Klabukov and I. V. Shtranikh

The data output devices for the measurement and registration center are time-shared and serve all information-gathering channels. On the one hand, therefore, such devices are used more compactly, and this is the great advantage of the MRC; on the other hand, they can be made with enhanced capabilities, thus reducing substantially the equipment expenditures for the center as a whole, while offering significant gains with regard to the output of data.

The following functional data input-output capabilities have been developed in the MRC system:

1) standard digital printout at a speed of 20 channels per second;
2) output to standard card punch at a speed of 10 channels per second;
3) output via cables to display oscilloscopes in digital form (univariate analysis);
4) output to a one-coordinate central-station recorder of the ÉPP-09 type and two-coordinate recorders of the PDS-021 type;
5) three-dimensional and color displays for the case of multivariate analysis;
6) back input of punched card data through a standard reading device of the ChU-KZMM-63 type at a speed of up to 20 channels per second;
7) output to off-line computer (such as the M-20 electronic digital computer).

Practically all of the devices constructed so far are transistorized. The networks rely heavily on trigger cells made up of tunnel-diode—triode combinations [1], as well as diode-transformer gates.

All of the data output and input devices are "synced in" with the operating cycle of the drum, i.e., with its arithmetic gate, channel, and half-turn pulse trains.

A block diagram of the data output and input network (except section 5) is shown in Fig. 1.

1. Digital Printout of Data

The printing device, which has a speed of 20 channels/sec, is so designed as to print any number in two stages; the first, or "active printout" stage takes 2/3 of a shaft revolution, or $2/60 = 0.0333$ sec; the second, or "waiting" stage takes about $1/60 = 0.0166$ sec, during which time the capacitors in the thyratron circuit of the printer electromagnets discharge. The start of each printout stage is properly timed by means of START PRINT and END PRINT pulses, which are generated in the printing device and are used to control the operation of the data-output-to-printer block.

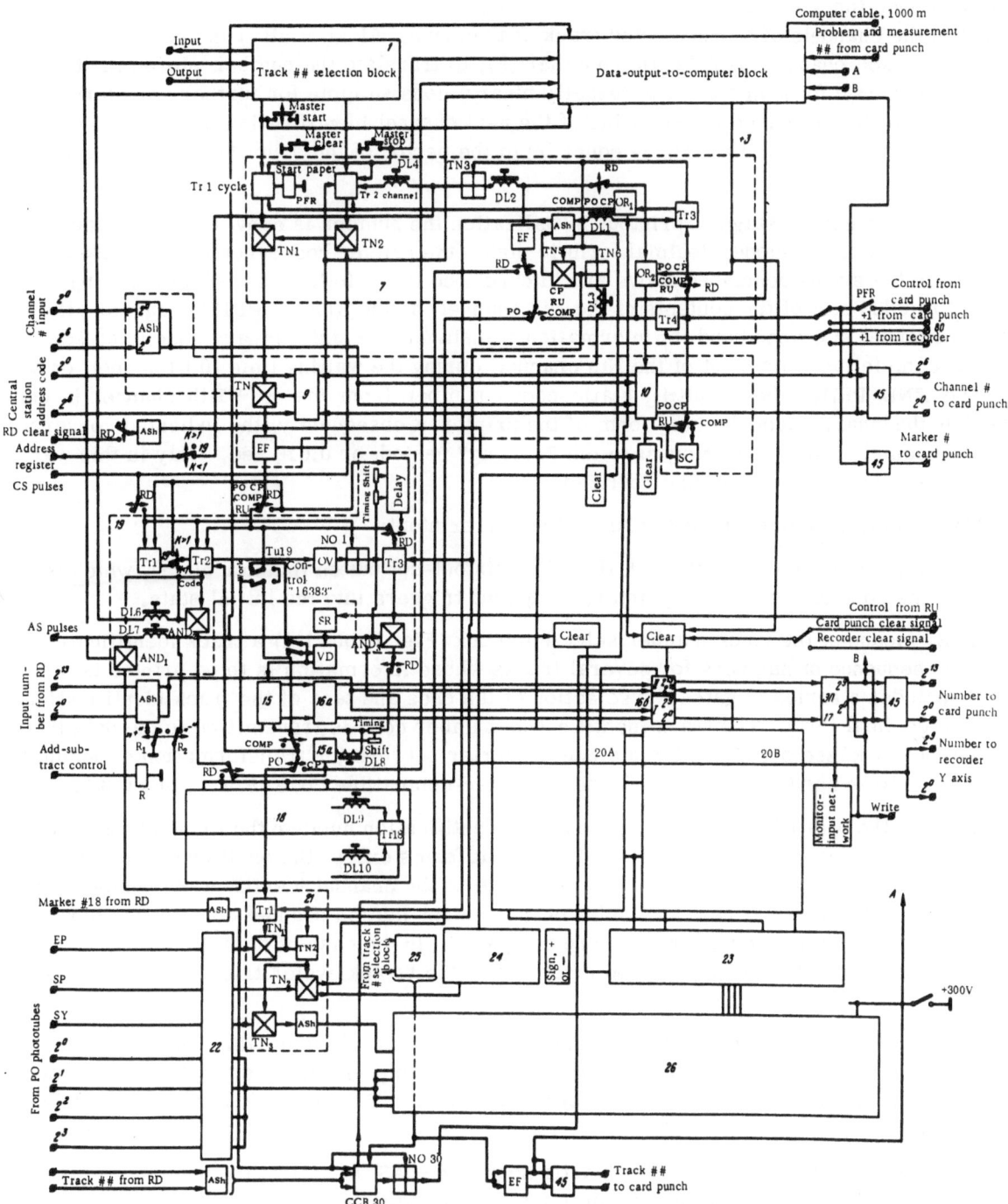

Fig. 1. General block diagram of the measurement and registration center data output network.
TN) Through-transmission network; ASh) amplifier-shaper; DL) delay line; UV) univibrator;
EF) emitter follower; SC) scaler; SR) scaling relay network for data output to recorder; VD)
variable delay based on stable univibrator; PO) digital printout; CP) card punch; RU) recording
unit; RD) reading device; AS) arithmetic pulse sequence; CS) channel sequence; PFR) paper-
feed relay; CCB) code comparison block; Tr) trigger; #, ##) identification number(s).

In connection with this, two techniques are used to set up for the printout of information carried in the channels of a preselected track of the drum. First of all it is necessary, after scanning the next channel identification number (#), to translate the number stored in that channel in dynamic form into a static (usually binary) code suitable for further transformation. The maximum time required for scanning of the next channel identification number and conversion of its contents into static form is equal [2] to the period of the drum "memory," or 1/50 = 0.02 sec.

During the second stage of printout initialization the number is converted from static binary code into binary-coded decimal form to match the printer code; then printout takes place. At this time the stages described above are interrelated in the following manner: During "active printout" (0.0333 sec) of the contents of the Nth channel, on the START PRINT command the (N + 1)th channel is scanned and its information translated into static binary code (0.02 sec). After the completion of printout of the Nth channel, during the waiting time (0.0166 sec), on the command END PRINT the resulting static-coded number from the (N + 1)th channel is converted into the binary-coded decimal form of the printer. Consequently, the indicated conversion must be accomplished in a time equal to the "waiting" period, or 0.016 sec. Only in this way can the maximum printout speed be realized.

The following main operations are realized in printout:

a) The printer motor is actuated, then the printout-set button (master clear), which is located on the control panel of the printer, brings all triggers into the initial state.

b) By means of the track number selection block (Fig. 1, block 1) a track identification number or sequence of numbers for several tracks whose information is to be printed is set. Block 1 includes a series of gates for selection of the ones and tens of the track identification number in decimal form. The track identification number code is transmitted to the data in-out switching network (see sec. 5) and to a diode converter network for transformation of decimal code into printer-, card punch- and off-line computer-compatible code (Fig. 1, block 25).

c) After the operation of scanning the track identification number the printout logic block 7 is actuated (this block is common to all data output forms except the oscilloscope display modes) and transmits gate pulses (channel sequence) to the channel identification number code comparison block (CCB) 9, where the codes of the central-station channel counter and binary channel identification number counter 10 are compared. In the case of digital printout, block 7 simultaneously actuates the paper-feed mechanism (the "track cycle" trigger Trl actuates relay PFR).

d) The phased-in code comparison pulse (comparison is initiated with code 000) in the form "+1" is sent to a 14-position shift register 15, beginning with the last position. This pulse concurrently deactivates the "channel trigger" Tr2, and gate TN2 of block 7 is closed to inhibit transmission of the gate sequence.

Then a train of timing pulses (arithmetic sequence — AS) is sent to the numeric register 15 through block 19. The same pulse train is transmitted through the delay line DL8 to one input of the phasing trigger 15a, whose other input receives the serially-coded number from the given channel, beginning with the last position, through the track identification number selection block 1 and again through block 19. The phasing trigger 15a controls the operation of block 16a, which after 14 shift cycles sequentially transmits the number from the given channel in reciprocal code to a binary 14-positon counter 16b.

e) The output pulse from the shift register 15 actuates the printout logic block 21, which remains on as long as information is being printed from all channels of the selected track. The printout control pulses START PRINT (SP), END PRINT (EP), and printout sync pulses Sy (gates) are transmitted to the inputs of gates TN1, TN2, and TN3 of block 21 from amplifiers 22, which amplify the signals from the photodiodes of the printer code drum.

f) The END PRINT pulse works through the opened gate TN1 to interrupt the numeric decade 23, initialize gates TN2 and TN3 of block 21, and then through gate TN6 (block 7), which is open in the initial state, to actuate the number transfer generator 20a. At this instant the information written in reciprocal code in the binary 14-place numeric counter 16b is converted into binary-coded decimal form, which is stored in the numeric decades 23 during the printout period of the given channel. The time allotted to conversion is 0.0166 sec, which would correspond to a transfer generator frequency of about 1 Mc. This is dictated by the fact that for a channel capacity of $2^{14} - 1 = 16,383$ bits, 16,384 pulses are needed for conversion (the overflow pulse in this case is used to deactivate the transfer generator, while simultaneously 16,383 pulses, i.e., one less than the number of pulses to the input of block 16b, is sent to the input of numeric decade 23). The inclusion of the simple transfer logic network 20B permits a sizable reduction in the frequency of the transfer generator (there are methods described in the literature [3] for reducing the conversion time at a diminished transfer generator frequency).

In the present article a simplified version is described, and a transfer generator with a frequency no higher than 100 kc is used. For the realization of this frequency the 14-place numeric counter block 16b is divided into two parts with two separate inputs such that a conversion network with a scale factor of 1024 by 16 is formed, where the arrival of one pulse in the second part of the network corresponds to the arrival of 1024 pulses at the input to the second part of the network. Similarly, in the case of conversion, in order to increment once in the second part of the network (Fig. 1, block 16b) it is required to transmit the code for the number 1024 into the numeric decades 23. These operations are carried out by the transfer logic network 20B. In the case in question the number of pulses in the conversion series is reduced to 1083 (at the input of numeric decades 23), which corresponds to a conversion time of 0.0108 sec.

After printout of one channel the START PRINT pulse adds "+1" to the channel identification number decades 24 and the binary address register 10, and the "channel trigger" of block 7 is actuated to scan the next channel of the selected drum track and to transfer its information in reciprocal code into the 14-position binary counter 16b during printout of the preceding channel identification number. The track identification number codes, channel number code, and "datum" code are transmitted to the printout matrices of the thyratron block 26. The printout sync pulses and codes from the printer code drum are sent to the other corresponding inputs of the printout matrix.

After conversion of the information from the 127th channel the START PRINT pulse increments the address counter by "+1," thus clearing it to "0," and the overflow signal flips the trigger of block 7, inhibiting the "channel trigger." Simultaneously the "track cycle trigger" of the printout block 7 is cut out. The next END PRINT pulse operates through the open gate TN5 of block 7 to cut out the printout logic block 21, clear the channel identification number decades, and return trigger Tr3 to its initial state.

For monitoring of the conversion subsystem it is sufficient to insert a monitor tumbler switch 19, which cuts out the shift register circuit 15, while the code comparison pulse is connected directly to the output of the printout logic block 21. If the blocks involved in conversion function properly, the maximum number is printed in each channel, i.e., 16,383. The actual printer is monitored by means of interchangeable fixed-program modules.

2. Data Output to Card Punch

This output mode is simpler than the printout mode in that it does not require conversion from binary to binary-coded decimal form. Card punching is executed in binary code except for the drum track identification number (#), which is punched in binary-coded decimal form. The information punched includes the contents of the channel (14 positions), the channel identification number (7 places), the control measurement symbols (2 columns), the track identification number (8 positions), the problem identification number (4 positions), and the measurement identification number (5 positions), adding up to a total of 40 binary positions (bits). In addition the following auxiliary symbols are punched: columns 17 and 18 — readout markers for the M-20 computer and MRC reading device, i.e., a symbol or command READ THIS LINE.

One channel is allotted to each line of the card, 12 channels per card. The standard operating speed is 50 cards/min, or 10 channels/sec. However, the time to punch one channel is somewhat less than 0.1 sec, because there are two "dummy" positions (lines) for transporting the cards. Consequently, the time per channel is about 0.086 sec.

After each punch in the dynamic output mode a pulse is transmitted from the card punch via bus 280 (see Fig. 1) as a signal to clear the numeric register 16b. Then an increment pulse "+1" is sent via another bus (80) to the binary channel identification number counter 10 through the circuits of block 7 common to both output modes.

The shaping trigger Tr4 of block 7 prevents the occurrence of several "+1" pulses in one operational cycle of the contacts of the card punch cam mechanism.

After admission of "+1" to the channel identification number counter, as for the digital printer, the identification number of that channel is scanned, and its contents are transmitted into the numeric register. This process requires a maximum possible time of 0.02 sec (the time for transmission of all 128 channels is equal to the drum memory cycle).

Then a punch control pulse (TRANSMIT NUMBER), which is the drive voltage (gate) for the servo output stages of the card punch electromagnets, is transmitted from the card punch via bus 278 to block 45, which controls the electromagnets for punching of the channel identification number, the number carried by the channel, and the track identification number. The problem number, measurement number, and control symbols K1, K2 are set up prior to the initiation of output, by means of a series of gates located in the card punch itself.

After the data of the 127th channel has been punched and "+1" transmitted to the address counter 10, as for the card punch, the resulting overflow pulse of the channel # binary counter 10 actuates the trigger Tr3, which precludes transmission through gate TN3 of the command to scan the next channel. Simultaneously the "track cycle" trigger Trl of block 7 is actuated, and the card punch start circuit is interrupted (the contacts of relay PFR are opened).

3. Output to Off-Line Oscilloscopes

The operating logic of the drum storage devices is readily amenable to the continuous oscilloscope-type display of all data, because the stored information is periodically read on all tracks by all heads simultaneously. The visual display of data stored on one track (univariate analysis) is very simple to realize and is a convenient form for the visual observation and monitoring of stored data during an experiment. For this purpose an accessory is attached to a typical oscillograph in order to generate a point raster (14 by 128 points) on the screen. A horizontal row of raster points corresponds to identical digital positions of numbers stored in all 128 channels of the selected drum track. A vertical row of raster points corresponds to all digital positions of the number in one channel (in our case we have a 14-place number), i.e., the observed spectrum is represented in digital (binary) form.

An oscilloscope accessory of this type is available to every user group. Three circuits are used to tie it in with the measurement center for display as described: a sync-mix pulse circuit (half-turn, channel, and arithmetic pulse sequences), a stored-information remote clear (erase) circuit, and an illumination pulse circuit (pulses from the output of the amplifier for reading of the displayed track). The selection of the drum track for visual display, as for the other output modes, is executed by the track identification number selection block, whose control-voltage outputs open the corresponding diode gate of the data input-output switchboard (Fig. 2).

From the master clear output for all tracks code pulses (display illumination pulses) for the selected track only are transmitted to the digital printout section for distribution into the desired output modes; visual display is never disconnected during any of the other output modes.

In our own case a type IO-4 oscilloscope is used; this instrument is recommended because on one of the posts of its back panel an ac voltage of 6.3 V is developed and can be rectified and used to supply the transistorized accessory circuits. The accessory is attached to the posts of the back panel where it does nothing to interfere with other applications of the instrument.

4. Output to Recording Units

This output requires some minor network modifications that affect only the given circuit, because all other circuits are common to all output modes and are designated as central-station networks. The modifications include a converter to transform the binary-coded identification number of any channel of the track into a voltage, and a scaling device to go from $1 \cdot 10^3$ to $16 \cdot 10^3$ pulses per scale.

For conversion ten of the fourteen positions of the number are used (the number of positions is dictated by the sensitivity threshold of the recording device). The converter operates on the voltage summation principle [4]. All resistances in the summing branches of the converter are chosen with 0.1% tolerances. The keys commuting the comparison voltage are transistorized and have a maximum remanent voltage of 0.05 V in the conducting state; this improves the conversion accuracy by a safe margin over the precision of the instrument itself. The gates are controlled from the output stages of the appropriate triggers (places from 2^0 to 2^9 inclusively) of the numeric register 16b (see Fig. 1).

The rate of information output is completely determined by the speed of the recording device. In our system we use an ÉPP-09 spool-type tape recorder and a PDS-021 two-coordinate recorder. The latter is recommended for its capability of superimposing several spectra at once for comparison. Since both devices have a one-second characteristic operating time, the output rate is about 1 channel/sec.

The same blocks are involved in the output of information as for the output to card punch, except that the circuits for clearing the numeric register 16b are switched over to the recorder-controlled mode by the addition of "+1" to the channel counter 10.

The device for change of scale has the following gradations with respect to the maximum stored number in the channel: 1, 2, 4, 8, and 16 (in thousands). Change of scale is executed by the scaling relay block SR (see Fig. 1), which is remotely controlled from the code-to-voltage conversion block. This block, in turn, controls the delay time of the univibrator (UV), whose function is to time delay the channel identification number code comparison pulse at the input of the 14-place shift register 15. This delay causes the binary register 16b to record a number in sequence, beginning with the position corresponding to the selected scale (delay), rather than with the last position. Thus, with SR in the setting 2 (two thousand), the transfer of a number into register 16b begins with second position, etc., and from this place on the numeric code is converted to a voltage.

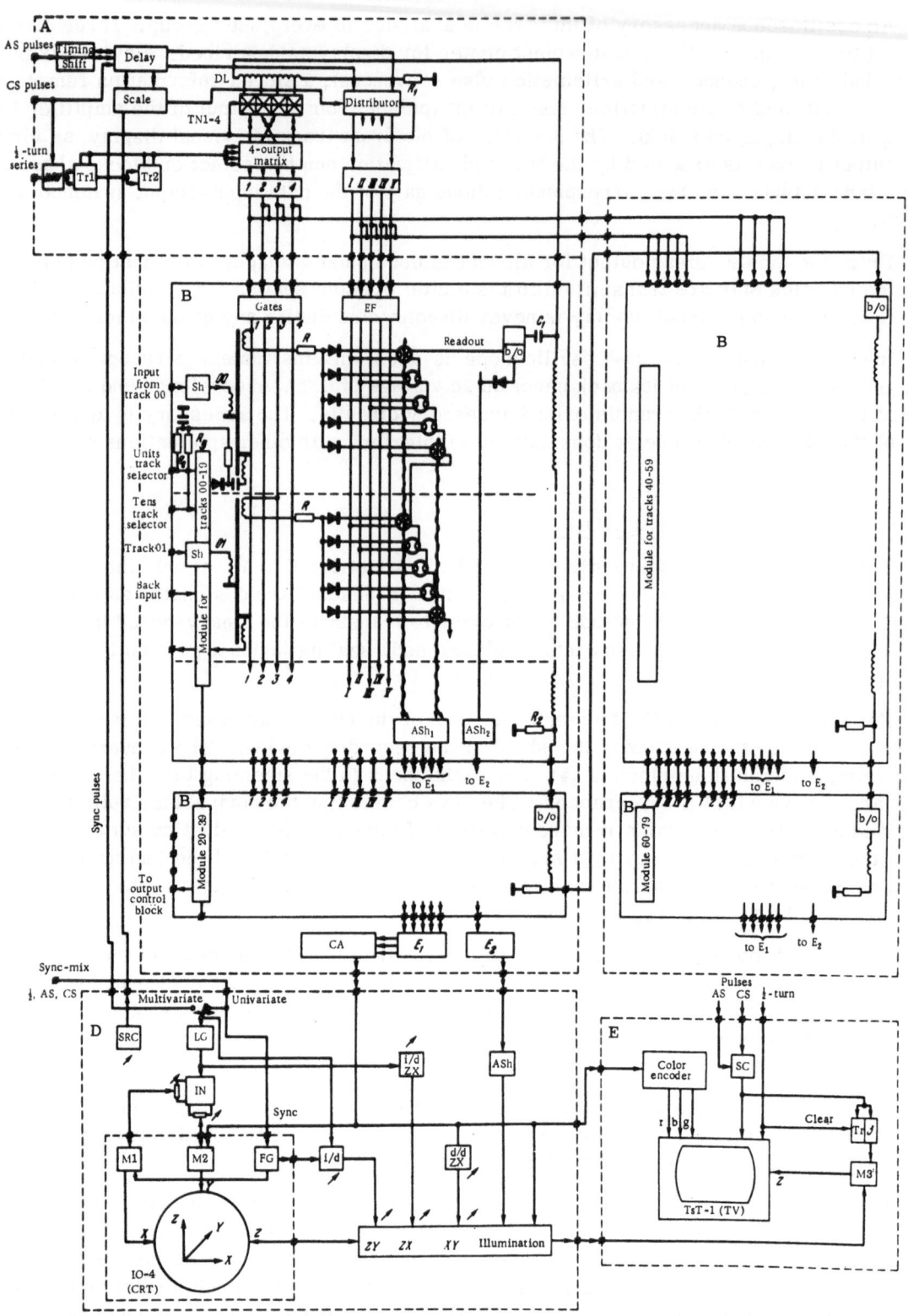

Fig. 2. Block diagram of the data input–output switchboard and blocks for three-dimensional display and color output for multivariate analysis. A) Visual output gate pulse block; B) data input–output switchboard and converter system for three-dimensional and color display; Sh) shaping stage; EF) emitter follower; ASh) amplifier–shaper; E) totalizer; C) code-to-analog

5. Three-Dimensional Display and Color Output

We now discuss in fuller detail the overall data output system of the measurement and registration system. This system comprises two independent functional subsystems: a switchboard for selection of the track indentification number for one-dimensional (univariate) readout to the digital printer, card punch, recording units, oscilloscopes, and cable-connected off-line computer, plus a subsystem for the multidimensional (multivariate) visual presentation of accumulated data in simple light-and-dark, isometric, or color display modes.

These subsystems share many elements in common.

Univariate Analysis. The circuits associated with the dynamic sequence of code pulses from the output of the channel accumulation block of each track are assembled in groups of twenty in one module (Fig. 2, block B). Each track circuit is terminated in a shaping stage Sh. The latter comprises one triode and a four-winding transformer, one of whose secondary windings is connected to a diode gate controlled through two resistors R_e and R_g by the gate bank of the track identification number selection block.

The only gate diode prepared for transfer is the one whose two resistors are supplied with a potential of −40 V. The combined output of the twenty-diode module is transmitted through a booster amplification network, then is connected to the other groups and fed into the digital printer housing for further distribution among the desired output modes.

Multivariate Analysis. In the case of multivariate analysis the output of data for visual display is complicated, on the one hand, by a sharp increase in the volume of information and, on the other, by the principal difficulty, which is the requirement for the creation of a relief picture, since the geometric representation of a two-dimensional spectrum is a three-dimensional surface. The multidimensional output problem is to present information in a form that graphically depicts the interrelationship of two (or more) given parameters such as X and Y. A number of techniques are available.

1) The illuminance of points on the XY-coordinate plane (for example) is made proportional to the values of the stored numbers. This technique is well-suited to fast subjective representation.

2) A second method entails the display of cross sections, with the individual spectra shifted along one or two coordinates on the screen (isometric display mode).

3) A third method is point-by-point display on a coordinate plane with the color characteristic of each point [5]. The color of the point of its spectral dependence is a function of the values of the stored data.

In all forms of this type of output undesirable flicker on the screen is avoided by keeping the frame rate on the screen above 25 to 50 cps. In the event half- or quarter-frames are used or interlacing is provided, the frame rate can be lower. In our work, in order to simplify the equipment and to facilitate the pulse selection and shaping conditions, we use four quarter-frames at a frame rate of about 50 cps, as dictated by the speed of rotation of the magnetic drum.

Even in this case, however, despite the use of interlacing, there is still considerable flicker. The latter is eliminated by the use, not only of longitudinal, but of transverse interlacing

system; D) three-dimensional display block; SRC) scale remote control; LG) line generator; IN) inverter; FG) frame generator; M) mixer; i/d) integral discriminator; d/d) differential discriminator; F) color output block; SC) scaler; r) red beam control for TsT-1 television set; b) blue beam control for television set; g) green beam control for television set; b/o) blocking oscillator; DL) delay line; Tr) trigger; AS) arithmetic sequence; CS) channel sequence.

as well. The points of each channel repeat every 4/50 sec, but at the same time the points on
the raster lines for adjacent channels are not selected just for the track identification numbers,
but are somewhat shifted in addition. This is done by switching to record from the tracks during each channel interval. In order for all tracks to be selected within the limits of one line in
4/50 sec, the counting network of the matrix triggers has 129 pulses, including 128 channel
pulses and one half-turn pulse.

For the writing of information on the drum surface when the data numbers and identification numbers of the channels are written serially, while channels carrying the same type of information in all channels are written in parallel, the following transformations are required: ·
conversion of the time-sequential codes of each datum into parallel codes and time sequencing
of the parallel output of information from like-numbered channels of different tracks. The
latter transformation must be executed in a period of time smaller than that required for a
change of channel identification number. There are several alternative methods for this conversion process. Historically speaking, the earliest and least auspicious for our purposes was
the method of equipping every track with its own "code-to-analog" network. These networks
were serially scanned in time (for example, by means of delay lines).

The implementation of this method in one laboratory disclosed the presence of difficulties
in the maintaining of identical operation on the part of all the code-to-analog converters.

A second technique, developed at FIAN and implemented in the same laboratory, consists
of sequential testing for the presence of a code pulse for each track. All tracks are parallel-
scanned for four or five pulses at a time. The code pulses of each track are transmitted
position-by-position through diodes to the appropriate locations of a special delay line. Identical positions share a common line, whereas the takeoff location is shifted in proportion to the
track identification number. By augmenting the output of the principal delay lines with secondary lines whose traversal times are larger, the lower the position of the principal line, it is
possible to obtain parallel output of the codes of each channel while the identical channels of all
tracks are taken in turn. The delay line outputs feed into a single code-to-analog conversion
network.

This method offers stable operation, but has two shortcomings: a) The number of delay
lines required is large (at 450 μsec with a pass band to 3 Mc); b) the period of the secondary
delay lines depends on the speed of rotation of the drum.

A third method has recently been implemented at FIAN, in which, as in the second method,
from each channel accumulation block data are taken in the form of pulse codes, but these codes
are transmitted to a matrix of ferrite cores (see Fig. 2). The core groups of this matrix are
then interrogated in sequence. This gives strictly time-parallel codes at the common output.

The multidimensional form of output comprises five blocks: A, B (four pieces), C, D, and F
(see Fig. 2). The gate pulse block A is designed, first, for selection of five adjacent out of fourteen
digital positions, so that 32 amplitude gradations are permitted. This block includes a scale
selection commutator in the form of a 9-place ferrite shift register with takeoff points (similar
to the numeric register of the printer). The register is controlled by the arithmetic gate
sequence of the read-amplifiers.

A remote-selected (from the oscilloscope block) position is transmitted to a second 5-
place register with five output buses, each of which carries a negative pulse shifted in time by
one arithmetic cycle. Pulse signals formed with a level from +20 V to zero are sent from the
distributor output to the multidimensional readout circuits of all eighty tracks, which are located in the two cabinets housing the logic blocks (four pieces in block B).

In addition to scale selection, block A has the second function of realizing the serial group writing and readout of data. If data from all eighty tracks are to be read out during each channel interval of about 156 μsec, it is required to maintain a separation of less than 2 μsec between them. In view of the necessity of added time loss for shaping, phasing, etc., as well as the possible use of lower-cost triodes, this time interval is doubled.

In turn, the use of a standard color television system operating at a rate of 50 half-frames per second and 312.5 lines per second requires that the time interval be cut in half, with the raster expanded to 256 lines. This is close to the indicated standard frequency and therefore does not demand any complex auxiliary regulation of the TsT-1 video system used in the center.

To meet the conditions described, block A has a read-write distributor comprising a matrix with four outputs. Each output circuit of the matrix gives a control voltage lasting about 156 μsec. In this way, during each channel interval-frame data are taken from one fourth of all the channels and are stored in the ferrite matrix. The distributor is controlled by two triggers, Tr1 and Tr2, which convert the channel pulses and drum half-turn pulses.

The final function of block A is to create alternating time delays on the part of the read pulses. This means that the point raster on the oscilloscope and color television screens contains 80 instead of 20 points for each of the 128 lines. For this purpose the pulse from the fifth position of the distributor, serving simultaneously as a synchronizing pulse for the scanning lines on the screen, is transmitted during each channel interval to only one of the four transmission networks TN1-4, which are connected to the delay line DL. At the DL output appear the read pulses, shifted in time relative to one another by intervals of from 0.9 to 2.7 μsec, thus creating a four-position deviation for each point of the raster.

The remaining winding of each transformer (Fig. 2, block B) can supply the write current through the resistors R and diodes to type K-272 ferrite cores with a rectangular hysteresis loop. The write current for each core is supplied only when the following three conditions are met:

1) The given group of write coils terminates in a connector of the distributor matrix without negative potential (−20 V).

2) One of the gate pulse lines (I-V) connected to the beginning of the transformer secondary of the shaping stage Sh carries a gate pulse instead of a potential of +20 V.

3) A read pulse for the given position must be present on the given transformer secondary. In this case the corresponding cores will be magnetically reversed.

After the fifth gate pulse a counting trigger pulse is generated and sent to the distributor delay line. The counting blocking oscillators b/o are triggered from the takeoff points of the delay line in 3.6-μsec intervals. The element used to connect the b/o trigger circuit and the line is the line capacitor C_1, which is connected to the base of a trigger triode with grounded emitter. This takeoff technique does not attenuate the signals in the line.

Each b/o generates a read current to an entire group of 20 cores. But only certain members of a subgroup of five can be reversed in that group (Fig. 2, block B), namely those which are tied in with one of the drum tracks and on which something has been written.

A group of common position buses penetrating all cores of a given block of 20 tracks will carry strictly parallel code signals. These signals are combined in a common circuit with all cores of blocks B and are transmitted to the code-to-analog network C (Fig. 2). This network is analogous to the converter network of the off-line recorder, except that only faster gates are contained in the comparison voltage commutation circuit. Tunnel diode triggers are connected into the second circuit of this network. This makes it possible to have the converted

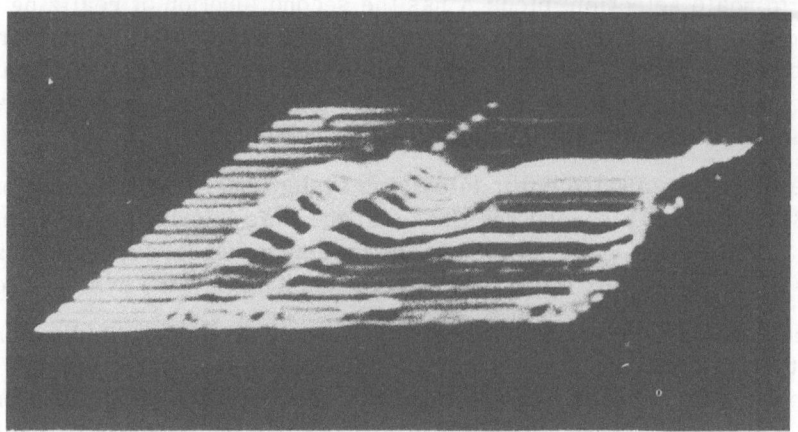

Fig. 3. Isometric display of a multidimensional spectrum.

code-to-analog value in the total commutation time for transfer from one track to the next. The resulting shape of the code-to-analog voltage is transmitted to the multidimensional display circuits.

Simple Light-and-Dark Display. A vertical raster is generated on the monitor screen. The track identification numbers are distributed along the vertical lines. The line numbers correspond to channel identification numbers. The modulation of the beam illuminance by the shape of the code-to-analog voltage presents a workable subjective picture of the output.

Isometric Display. By slanting the raster and sending the code-to-analog voltage to the amplifier of the vertical-deflection plates it is possible to obtain an isometric representation of the observable spectrum, as shown in Fig. 3. At the same time, by increasing the beam illuminance in proportion to its degree of vertical deflection it is possible to further enhance the effect.

This kind of display permits the introduction of reference marks for the cross section of the spectrum. The output system (Fig. 2, block D) has two integral discriminators i/d (operating on the level of increase of the line and frame deviations) and one differential discriminator d/d (operating on identical output level), which isolate, so to speak, isonumeric cross sections.

Color Output of Data. By transforming the instantaneous values (Fig. 2, block F) of the code-to-analog voltages into three functions of the type shown in Fig. 4 it is possible to produce a spectral scale on the screen of a three-beam mask-plate tube. A raster similar to 1a permits the data output to be represented as a conventional geographical color-topographical map. The number of distinct gradations can be as high as ten, i.e., in principle it is possible by this method to observe with 10% error as many as 500 thousand channels on a standard television screen (with a raster of 600 lines).

The conversion block for transformation of output level into three color-differentiated components comprises a bank of diode matrices and simple linear gating networks (see Fig. 4). All multidimensional forms of data output have a dot-array pattern determined by the illumination gate pulse sequence, which is generated by the bank of counting blocking oscillators of blocks B.

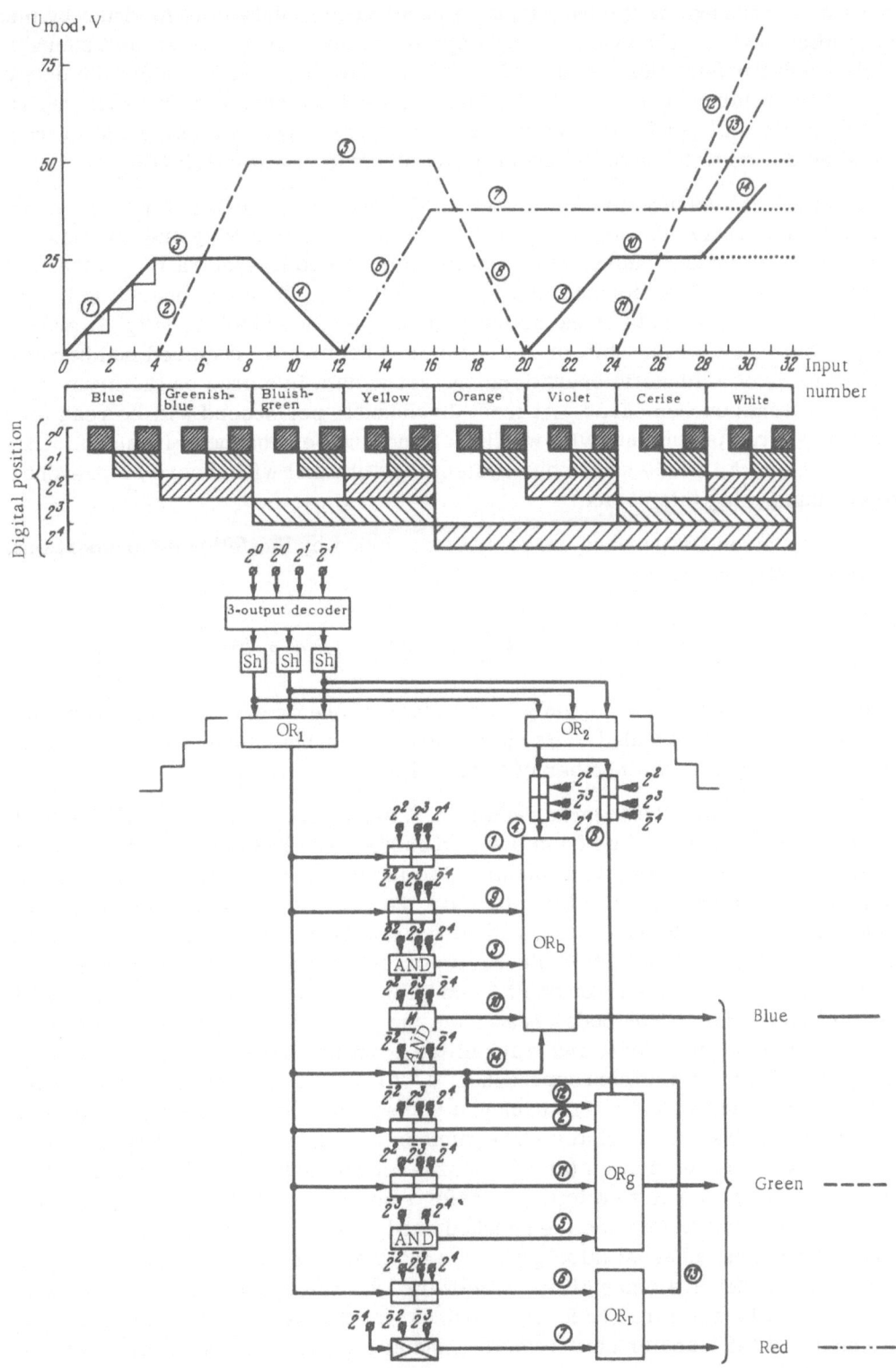

Fig. 4. Characteristics and block diagram of the color encoder.

6. Data Input from Punched Cards

A facility is proposed for the back input of punched-card data into the drum by means of a standard reading device. The channel identification number codes are simultaneously read and stored (through the amplifiers of block 8) in the address counter 10, while the data code and track identification number code, marker No. 18, are read and stored in numeric register 16b (through the amplifiers of block 14). Before these codes are read, a clear pulse is transmitted from the reading device to the address counter and numeric register 16b (Fig. 1).

Often in physical measurements it is required to execute simple arithmetic operations such as channel-by-channel addition of certain data to other data having any coefficient and sign (for subtraction of normalized background, subtraction of certain spectra from others, etc.). These operations are easily performed in a system with logic elements operating in dynamic binary code. The device consists of the actual logical operation block 18 (Fig. 1) and a delay block 19, which permits a single shift of the registered data by the required number of places. For the performance of arithmetic operations one set of data is stored on the drum, while the other set is read from punched cards. At least 20 channels per second will be read for a reading rate up to 100 cards per minute with writing at one line per channel. Inasmuch as the access time to drum "memory" is 1/50 sec in our system, any channel will always be present on the drum at some time during this period.

The data written on the punched cards must be varied by the following proportionality factor, common to all channels:

$$K = k_1 \frac{1}{2^n} + k_2 \frac{1}{2^{n-1}} + \cdots + \frac{1}{2^0} k_n + p_1 2^0 + \cdots + p_m 2^m + \cdots ,$$

where the subcoefficients k_i and p_i can only have values of one or zero. Punched-card data, multiplied by the factor K, can be added to or subtracted from drum-stored data. The number of holes in the cards is equal to the number of terms of K.

We first of all examine the logic for addition of two numbers, one of which is read from the drum in dynamic form, the other from punched cards. The read marker No. 18 characterizing the command READ THIS LINE is transmitted to the track identification number comparison network CCB 30. Also transmitted to the latter are the track identification number pulses read from the punched cards and the static track identification number voltage levels from the diode matrix controlled by the track selection block. If correspondence occurs, the CCB 30 output pulse opens the channel sequence through TN1 and TN2 of block 7 to the channel identification number comparison network. The codes read from the punched cards, written in the address counter through the read amplifiers, and transmitted from the central-station address counter transmit a pulse at the instant of correspondence to Tr1 or Tr2, as well as to the network of the coefficient block. Triggering of Tr3 causes a pulse sequence to be sent through AND_3 of block 19 to the number transfer register. The number read from a punched card is written in parallel in direct code for summation (and in inverse code for subtraction) on the binary counter 16b. When a timing sequence is sent to the transfer register, its output develops a code in dynamic form for the number written on the punched card, synchronously with the code for the number from a given drum channel (also gating the arithmetic sequence). Beginning with the last positions, these codes are transmitted to network 18. The operation of this network can involve eight combinations of three concurrent pulses: at the output of the amplifier from the drum, at the output of the transfer shift register, and at the output of the transfer circuit for operation of trigger Tr18.

Number transfer is realized by 5-μsec delayed switching of trigger Tr18, which is switched by the nearest arithmetic marker, this event serving for the writing of the next placè. Since the counting amplifier is gated with a time lead (after suitable placement of the read head) of four or five microseconds relative to the corresponding write gate pulse, it is only necessary in the channel logic network to set the phasing trigger Tr2 in the required state prior to writing [2], as in the operation of the write logic network. For this purpose a control pulse, its polarity depending on the results of the network logic operation, is transmitted from the network output to the logic network of the given drum track.

For the subtraction of numbers from drum-stored data the inverse number is written in binary counter 16b, and to compensate for the "one" lost in this operation trigger Tr18 is switched on prior to the beginning of the logic cycle, generating "transfer" or "+1."

To obtain the coefficient k for the number read from the cards, it is necessary to shift the relative digital positions of the two numbers. With a shift of the number from the register to the right (time delay) K is increased by 2, 4, . . . , etc., to as high as 2^{14}, and with a shift to the left (time lead) K decreases, becoming equal to 0.5, 0.25, . . . , etc., down to $1/2^{14}$, all other logic operations remaining the same. These shifts are executed by network 19. For the operation of delaying the code pulses obtained from the transfer shift register (K > 1) the signal from the network for comparison of the identification number of the required channel is written in "+1" form in the auxiliary shift register 19, which functions as a delay line. A bank of tumbler switches makes it possible to pick up the delayed transmitted pulse at any time from zero to 14 timing cycles. This pulse switches on the trigger Tr3; then an arithmetic sequence is sent through the AND_3 gate to the main register for transfer of the number picked up from the punched card.

For the operation of advancing the code sequence written in the binary counter from a punched card, the channel identification number scanning address counter is incremented each time by one, so that the comparison signal occurs one channel earlier. All other operations are the same as before. Through AND_1 and AND_2 trigger Tr2 always controls transmission of and read signals and controls transmission of signals for control of the logic trigger of the track circuit only during the time interval allotted the given channel on the drum. During operation with time lead the trigger Tr1 is switched on one channel earlier and, on being switched off, switches on Tr2.

In combination with the AND_3 network, the univibrator UV, which has a holding time greater than the channel period, prevents premature deactivation of Tr3 of block 19, so that the number transfer shift register is always completely clear. The process of scanning and switching of triggers Tr1 and Tr2 of block 7 from the track selection block must be terminated before the read pulses are received from the next channel; for this several blank cards are inserted in the interims.

In the event of a mismatch between the track identification number read and the one actually set, a signal appears at the output of the NO 30 network, which shuts off triggers Tr1 and Tr2 of block 7 through OR_1, deactivating the channel identification number code comparison block. In addition, the reading device start circuit is shut off, generating a malfunction signal.

7. Output to Off-Line Computer

The extremely large volume of digital material gathered in the registration of multivariate spectra to be computer-processed makes it imperative to have efficient communication between the measurement center and the off-line computer.

Indeed, if a relatively small 128 × 32 spectrum, i.e., 4096 channels, is to be registered, its recording (for input to computer from punched cards) calls for about 350 cards (punching one channel per line) and lasts more than 6 min (at a speed of 10 channels/sec), and machine input process requires even more time. Inasmuch as data must be taken five to ten times in the course of registering one such spectrum, the need for the incorporation of other interface techniques becomes manifest.

In our case direct communication between the MRC and off-line computer is effected via a single coaxial cable about 1 km in length. Data transfer is by serial code signals. In the first actual interface system, in order to simplify it, we employed direct writing of each number into the machine register R1, which was used as a shift register. The data for each number are transferred from R1 into magnetic operational memory, from which whole sets of numbers are then rewritten on magnetic tape of the computer. The transfer time for the spectrum indicated above into operational memory should take a total of about 2 sec.

As mentioned, the registration of a spectrum is accomplished in channel groups. Each group comprises 128 channels with 14 digital positions each, allocated to one track of the drum. The data interface operation could be limited to a total "word" of 21 places for each number: 14 places for the actual data number and 7 places for the channel identification number associated with that number, preceding each group with its identification number — 8 places (4 × 2). However, for increased reliability and the preservation of identity with recording on punched cards, the track identification number is transmitted with each channel of the group.

Besides these 29 positions, it is necessary by analogy with the punched-card record to transmit the problem identification number (4 positions), the identification number of the next measurement (5 positions), and the control spectrum identification number (2 positions). Thus, the total word comprises 40 places (Fig. 5).

For control of the interface process each word terminates in an end signal a, which represents a command to rewrite into computer operational memory. Since transfer is executed asynchronously relative to the computer cycle, after each position of the work (positive pulse polarity), regardless of whether or not it occurs, a sync pulse b is transmitted as a shift command to the machine register R1.

For initialization of the input network of the computer prior to the beginning of transfer from the center and for disconnecting it on completion of transfer, the above-mentioned pulses are accompanied along the cable with a code combination c corresponding to the beginning of transfer and a combination d corresponding to the end of transfer. The process of subsequent transfer into machine operational memory will be discussed in another paper.

As mentioned earlier, in the MRC the 14 positions of each number are placed in sequence on the drum track. Each track comprises 128 numbers, which we call channels. Inasmuch as the transfer of 40 positions is executed with the timing frequency of the arithmetic sequence of digital positions, the contents of every third channel are transmitted, and the entire transfer of information from one track takes three half-turns of the drum, or 3/50 sec.

The transfer of a complete word is executed by means of the bank 1 of diode-transformer gates M (see Fig. 5), which are connected on one side to the potential circuits from the arithmetic and address registers of the digital printer, the output circuits from the track identification number matrix, etc. To the other side of these networks is sent an interrogating pulse from the appropriate positions of the transfer shift register, which is incremented each time by "+1." All of the AND networks are interrogated in turn at the timing frequency, which is equal to the frequency of the arithmetic sequence. The above-indicated order of transfer of the channels 0-3-6, etc., is realized by the input of three pulses instead of one to the address counter each time ("+3" operation).

In order to obtain the compact output mode discussed, the entry of "+3" into the address counter, the command to scan the identification number of the next channel, and the extraction of its numerical datum are executed the instant that transfer of the preceding datum and corresponding channel identification number code is completed. The three pulses used for "+3" operation are taken from definite cells of the computer interface register (see Fig. 5), with two of them being transferred directly to the input of the channel identification number address counter

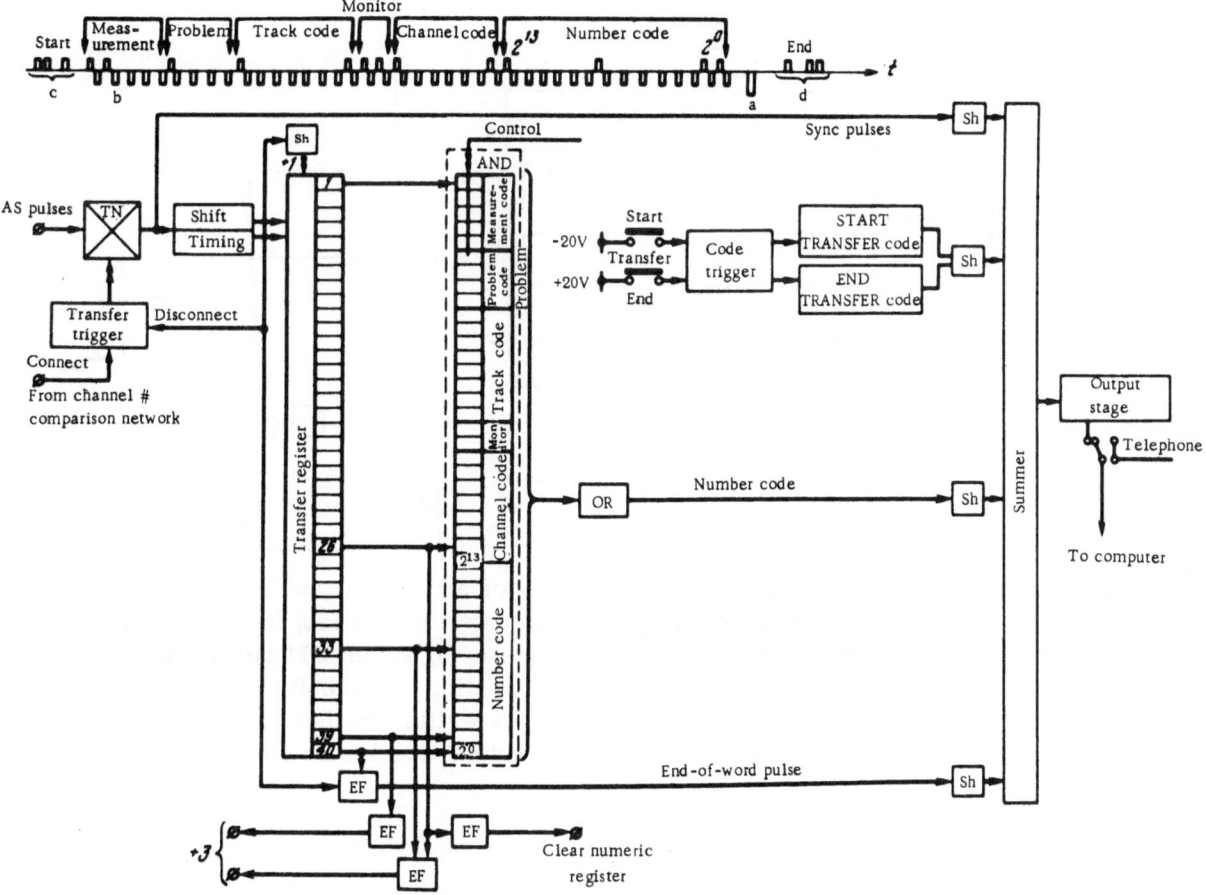

Fig. 5. Block diagram of the off-line computer information interface subsystem.

(through the internal commutation circuits of the latter), while the third, chronologically the last, is transmitted to the logic block 7 (see Fig. 1), which is common to all output modes, where it controls the block as in the "+1" operation for data output to punched cards.

Since the output of information from one track is interrupted with the onset of an overflow pulse from the channel identification number address counter, in the case of data transfer to computer an auxiliary three-way divider is connected to the output of the address counter, because in "+3" operation we now have three overflow pulses.

Figure 5 presents a block diagram of the subsystem for interface with an off-line computer; the subsystem contains a 40-cell transfer register and as many diode-transformer AND gates. The transfer trigger, which consists of a tunnel diode and triode, controls the transmission network TN and initiates transmission of the pulses of the arithmetic sequence (AS) to the stages for shaping of the shift and timing pulses to the transfer register and machine register R1. It is controlled by pulses from the output of the channel identification number comparison network, which signal the beginning of information transfer from a particular channel, and by pulses from the output of the transfer register, which correspond to the end of transfer of a word. The code pulses, timing pulses, and pulses from the output of the transfer register pass through additional shaping circuits and enter the master totalizing stage of the interface control block.

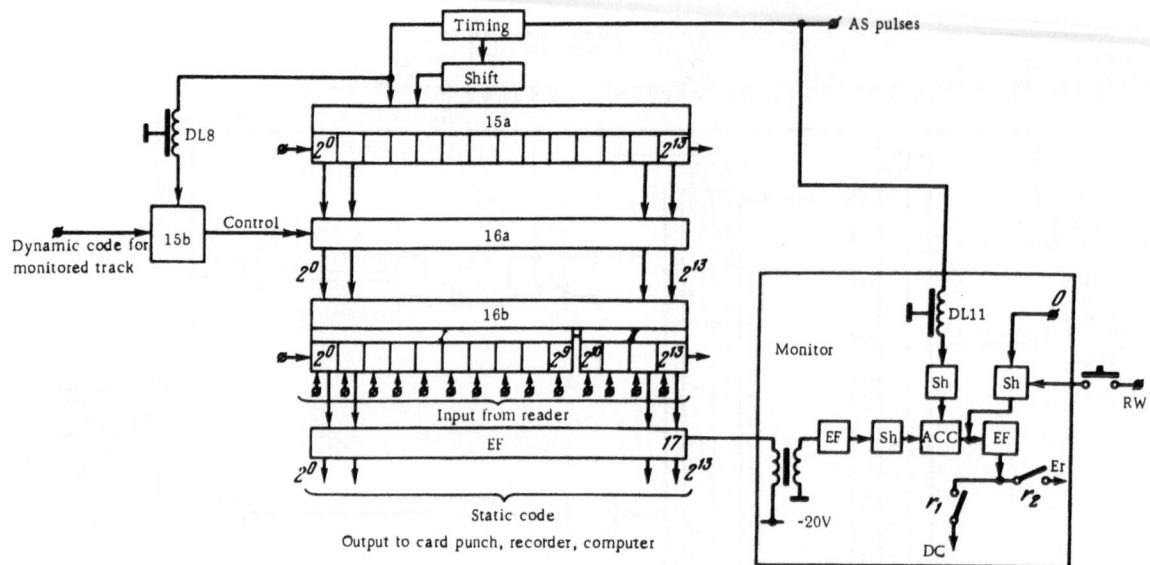

Fig. 6. Block diagram of the digital data output monitor and back input from the reading device. 15a) Shift register; 15b) phasing trigger; 16a) control gates; 16b) binary counter; EF) emitter follower; AS) arithmetic pulse sequence; Sh) shaping stage; DL) delay line; ACC) anticoincidence circuit; DC) dynamic code output for writing onto monitor track during data output, input from reader, rewriting from any track, or writing by light-pen modification of display; O) from light-pen amplifier; RW) rewrite from any track onto control track; Er) light-pen erasing.

This block has, in addition to the totalizer, a code trigger, START TRANSFER and END TRANSFER code generation stages, stages for shaping of the pulses taken from the transfer register, and an output totalizer stage operating into a cable of length ~ 1 km. When transfer is not in progress, the cable is used for direct telephone communications between the center and the computer facility.

8. Data Output Monitoring and Input from the Reading Device

The centralization of information measurement and readout presupposes the existence of a versatile facility for monitoring the operation of the main subsystems. This is particularly urgent with regard to those devices of the MRC which serve several different user groups in common. Among such devices are the main memory, or magnetic drum, and the data output subsystems.

The possible monitor modes are diverse. Inasmuch as the type of measurement center in mind is essentially a special-purpose "fixed-program" computer, monitoring is realized by the "wiring-in" method, which has the principal advantage of keeping the monitoring operations concurrent with the main operations.

We now examine in detail our actual system for monitoring the operation of the data output devices. First of all, monitoring is delegated to a subsystem that participates in all digital information readout modes (digital printout, card punch, off-line recorder, output to off-line computer). This subsystem executes the function of converting the dynamic sequential binary code (14-position) registered in any of the 128 channels of any of the 80 tracks on drum memory into static binary code.

Fig. 7. Test pattern for monitoring the performance of the reading device.
1) "TEST"; 2) "MRC."

The subsystem consists of a shift register 15a (Fig. 6); gate circuits 16a, which are controlled by the phasing trigger 15b; the binary counter 16b; and output stages 17. The operation of conversion to static code is effected in time sequence during 14 timing cycles (to match the number of places in the number code) and has been considered in detail in Sec. 1. The dynamic monitor code is obtained by connecting into the common collector circuit of all the output stages 17 an auxiliary differential transformer, whose output develops a time-serial back monitor code. The pulses for this code are shaped and transmitted to the anticoincidence circuit ACC, whose other input receives the 14 pulses of the delayed arithmetic sequence, so that from the ACC output are taken direct monitor code pulses, which are written onto a special monitor track.

The indicated writing operation is realized connecting the channel storage logic trigger to two additional independent start circuits, from which a signal causes the logic trigger to go over either into the WRITE position or into the ERASE position. Writing of the monitor codes in the form of illuminance pulses produces a signal that is transmitted to the monitor oscilloscope, where the monitored (readout) spectrum is on continuous display and an additional channel-by-channel "color trace" of the monitored spectrum is generated. Perfect matching of the "color trace" indicates proper operation of the monitored subsystem. The network is simple and contains few elements. It is important to note that this auxiliary network not only performs an output-monitoring function, it also serves without any modification for the reverse input of data from punched cards by means of the reading device. For this purpose the direct number code read from punched cards is sent in parallel form directly to the binary counter 16b (see Fig. 6). Simultaneously the channel identification number code is transferred to the address counter (see Fig. 1, block 10). After these codes are registered, scanning for the corresponding address code of the position on the monitor track of the magnetic drum is initiated. When an address code comparison signal occurs, the subsequent operation of the subsystems is the same as in the monitoring of data output. The only difference is that during input from the reading device the data number codes are not transmitted to the phasing trigger 15b. Special test patterns are used to monitor the reading device (Fig. 7). The presence of a reverse input capability permits (in addition to simple arithmetic operations) the correction of operator error during the setup of ancillary indices of the experiment, such as the measurement identification number, problem identification number, etc.

For the correction of visible errors in the main information a "light pen" is used (this is one of its possible uses). The light pen comprises a plexiglass cone-tipped holder, a flexible fiber-optic light conductor, an FÉU-54 photomultiplier, and a preamplifier with pulse time shaping.

A point raster of 128 × 14 points is continuously regenerated on the monitor screen. Each point of the raster corresponds to a definite time-position (digital place of the number from a given channel) on the monitor track of the drum. The brightness of the point, on the other hand, indicates the presence of information in the given digital position (of 14 positions) of the given channel (of 128 channels). When the tip of the light pen is applied to any point of the raster, a signal is generated, which on command from the operator causes either missing information to be written in or superfluous or erroneous information to be erased.

The measurement data to be corrected are written onto the monitor track either by means of the reading device or merely by depression of an auxiliary button marked REWRITE (see Fig. 6).

In the latter case the information from any track of interest at the moment is transmitted in pulse form through the closed contact of the button and the shaping stage Sh to the output stage of the monitor network. After rewriting, "correcting" pulses are transmitted to the same network from the light pen and, depending on the command (erase or write), act through the corresponding contacts (r_1, r_2) of an auxiliary relay to control the storage logic trigger, executing either write-in of the missing position in the number of any channel or erasure of a superfluous position.

LITERATURE CITED

1. V. V. Puzanov and I. V. Shtranikh, Trudy FIAN, 42:39 (1968). [This volume, p. 39.]
2. A. M. Klabukov and I. V. Shtranikh, Trudy FIAN, 42:53 (1968). [This volume, p. 8.]
3. L. A. Matalin, A. M. Shimanskii, and S. I. Chubarov, Advanced Scientific, Engineering, and Industrial Practices, No. 5, TsITÉIN, Moscow (1961).
4. É. I. Gitis, Information Converters for Electronic Digital Computer Equipment, Gosénergoizdat, Moscow (1961).
5. I. V. Shtranikh, V. N. Bochkarev, A. N. Volkov, V. M. Geraseev, A. M. Klabukov, V. V. Puzanov, and A. M. Shimanskii [Shimansky], Internat. Symp. Nuclear Electronics, Paris (1963), p. 587.

SUBSYSTEM FOR READOUT OF THE
INSTANTANEOUS VALUES OF THE TIME CODE

V. V. Puzanov and I. V. Shtranikh

In multichannel time-base spectrometers, which normally operate in the off-duty mode, the access time to main memory becomes important. If this time is greater than the channel on-time, distortions can occur in the registered spectrum [1]. Moreover, such systems require special circuit strategies in order for the counting circuits to be able to "catch up" in the time allotted to writing in the given channel. For spectrometers of this type a wise approach is to use prestorage buffer devices of the cumulative or derandomizing type [1-2]. In buffer systems utilizing half-current ferrites [1] difficulties arise in connection with commutation of the currents and the creation of proper current-holding conditions.

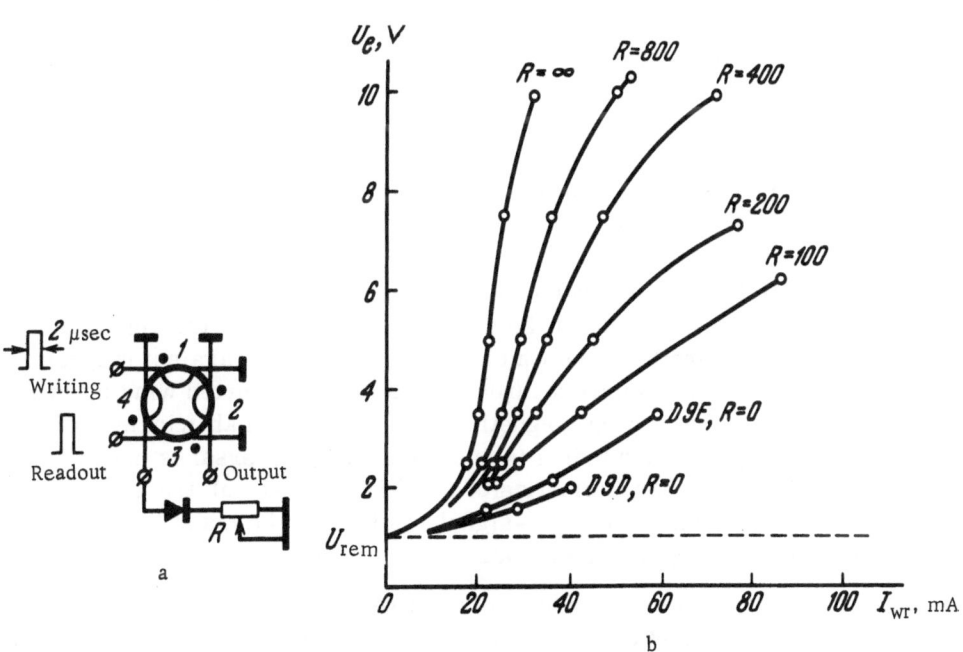

Fig. 1. Schematic diagram of the network controlling the writing of information into a core of material 0.25VT, $4.10 \times 10 \times 15 \times 30$, with a rectangular hysteresis loop (a) and characteristics of the false-one signal amplitude (b). U_{rem}) Signal voltage at $I_{wr} = 0$ with the coil closed.

27

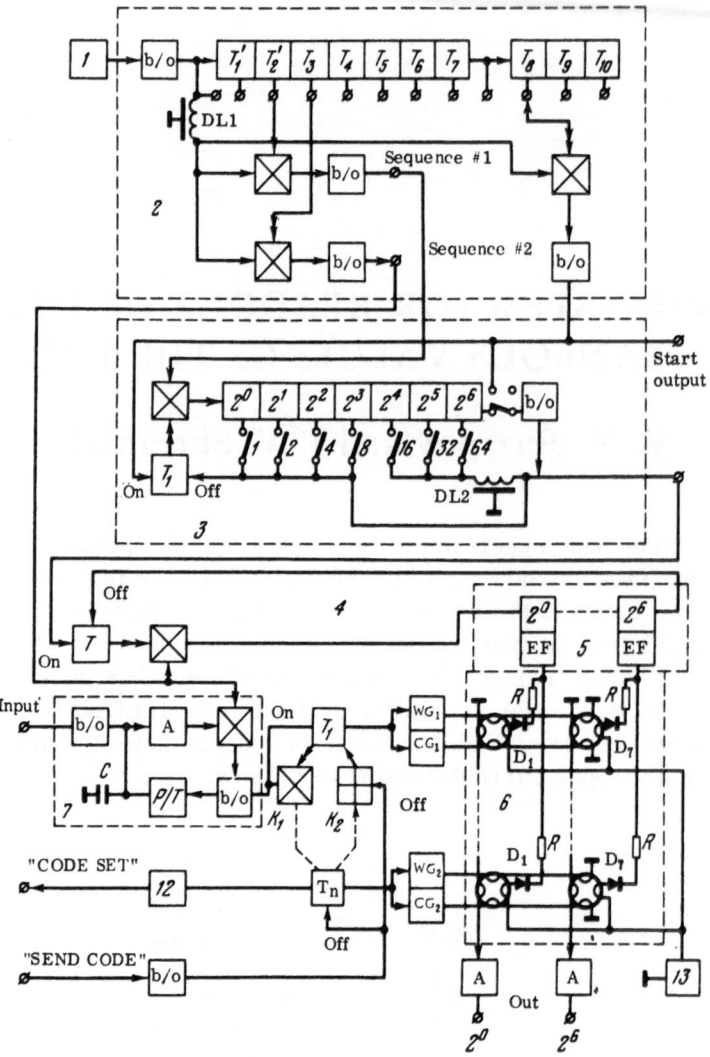

Fig. 2. Block diagram of the subsystem for readout of the
instantaneous values of the time code.

As a part of the MRC we have developed and tested a subsystem that is devoid of these
shortcomings. The operating principle of the subsystem is based on the generation of an oppos-
ing magnetic field for those ferrite cores which are to be inhibited against magnetic reversal
in the writing process, by sending current through an auxiliary shorted winding, which demag-
netizes the core by virtue of the strong magnetic coupling between all windings on the vertical
parts of the hysteresis loop. If the winding is open, magnetic reversal is not inhibited. Total-
current counting is used. The characteristics of the false-one signal are given in Fig. 1b. The
ratio of the ones-counting signal to the zero-signal reading can be made higher than ten, depending on the degree of shunting of the auxiliary inhibit winding.

A complete block diagram of the subsystem is presented in Fig. 2. It includes the master
pulse generator 1, start-pulse and gate-pulse generator 2 with phased output, time-delay network 3 with start-pulse generator 4, and the above-mentioned bank of measurement triggers 5
in combination with an "n-word" ferrite matrix 6 and input pulse phasing network 7. In addition,
there is a distributing network comprising triggers T_1, T_n, etc., through-transmission networks
K_1, K_2, etc., as well as read and write generators WG_1, RG_1, WG_2, RG_2, etc.

The pulse to be written is phased by network 7 so that it occurs at a time between the channel sequence pulses at the input of the triggers 5 (allowing for the firing delay of the latter). The trigger T_1 is fired, the leading edge of its pulse actuating the generator WG_1, Simultaneously trigger T_1 energizes the start circuit of trigger T_n, whereupon the second incoming pulse for writing fires the trigger T_n, the latter actuates the generator WG_2, and so on, and as a result the counting cells 5 are taken into the proper state for magnetization of certain cores.

Cophased read pulses arriving from the main memory network switch the triggers $T_n, \ldots, T_1$ in reverse order. The trailing edge of their pulse (the leading edge relative to the time of the read pulse) controls the read generators RG. The command to send read pulses is supplied from main memory in the event at least one of the triggers $T_1, \ldots, T_n$ goes over into the write state, thereby transmitting the CODE SET voltage level through module 12 to the main memory system. The entire subsystem is made up of modular semiconductor elements (six types). Several modified counting cells, of the types described in [3], are used. A 100-kc quartz oscillator [3] with inclusion of an emitter follower is used as the master source. The temperature stability of the oscillator is better than 10^{-6} per degree. Type K-272 cores, $4 \times 2.5 \times 1.5$, are used in the ferrite matrix of the buffer memory. The diodes $D_1, \ldots, D_7$ are wired into the circuit of the inhibit windings, where they exclude the influence of the latter during reading, along with influence of certain cores on other cores during writing. The resistances $R_1, \ldots, R_7$ safeguard against the passage of excessive control currents between the base emitter follower and the cell followers during the write-inhibit state.

The principle described here can be realized in very high-speed systems with use of small-diameter cores.

LITERATURE CITED

1.   L. D. Matalin, Advanced Scientific, Engineering, and Industrial Practices, No. 5, TsITÉIN, Moscow (1961).
2.   V. V. Puzanov, I. V. Shtranikh, and D. T. Matachun, Pribory i Tekh. Éksperim., No. 3, p. 82 (1966).
3.   S. Schwartz (ed.), Selected Semiconductor Circuits Handbook, Wiley, New York (1960).

A SIMPLE PERFORMANCE-CURVE TRACER

V. V. Puzanov and I. V. Shtranikh

The proposed performance-curve tracing network (Fig. 1) represents a modification with considerable simplification of an instrument designed to record the static characteristics of semiconductor devices such as conventional and tunnel diodes, p-n-p and n-p-n transistor triodes, and other elements. In the testing of triodes a family of two performance curves is displayed on the screen of an ÉO-7 or similar oscilloscope (Fig. 2a), showing the dependence of the collector current on the collector voltage as the latter is regulated from zero to 40 V for two values of the base current. These values may be varied in pairs, setting the following gradations: 0-50, 100-200, 300-400, 500-600 μA, as well as 0 separately, or regulated smoothly from zero to 800 μA. Diodes are tested with a regulated current from 0 to 40 μA or, when the button K_1 is pressed and the tumbler switch T_2 is turned on, with a regulated reverse voltage from 100 to 300 V.

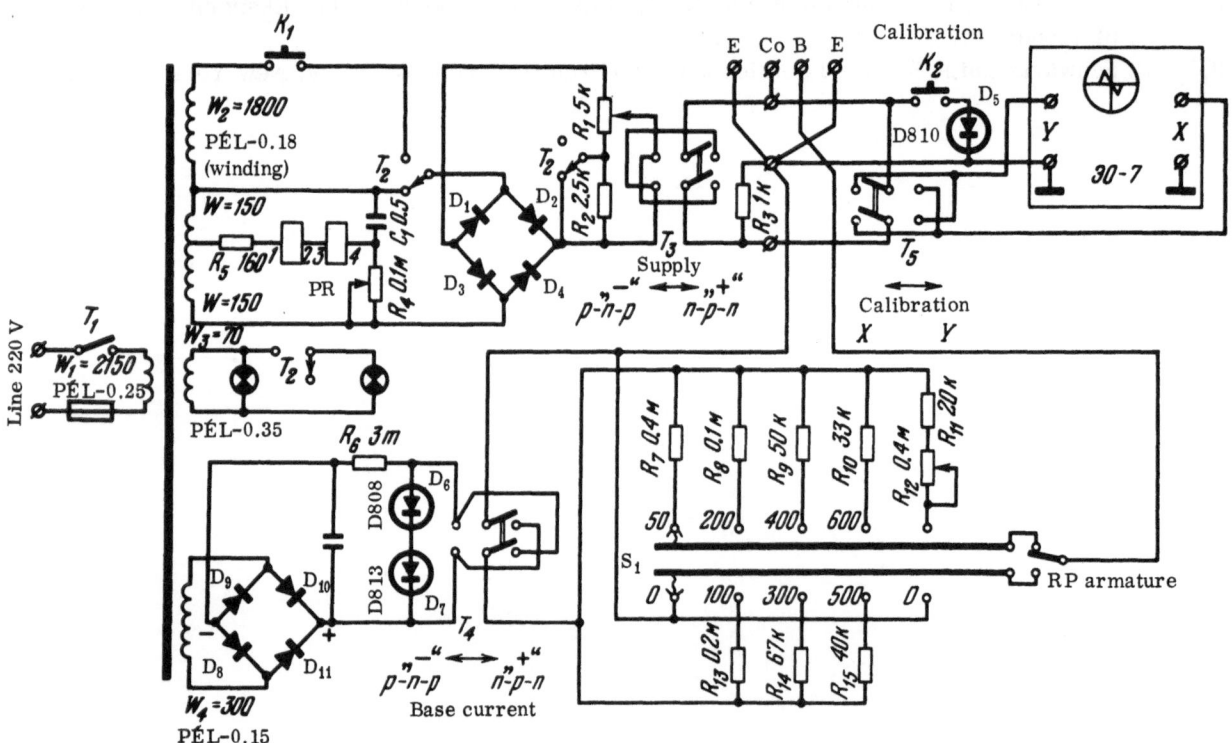

Fig. 1. Schematic of the performance-curve tracer.

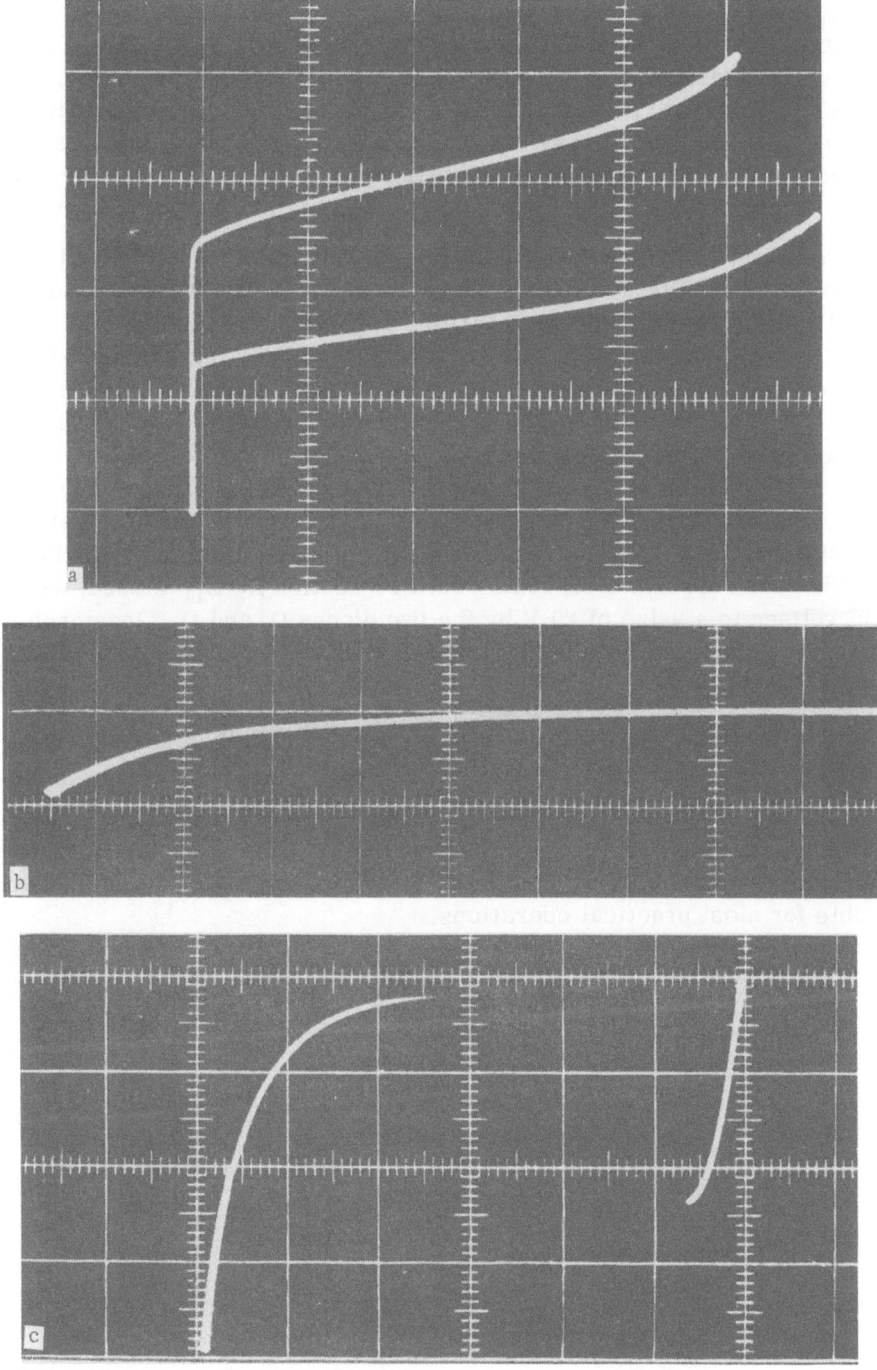

Fig. 2. Performance curves of D9K and R2 diodes and an MP42 triode. a) Family of collector curves for the MP42 triode at I_{b1} = 50 μA and I_{b2} = 100 μA (X-axis scale: 1 square = 5V, Y-axis scale: 1 square = 2 mA); b) reverse volt-ampere characteristic of D9K diode (X-axis scale: 1 square = 5V, Y-axis scale: 1 square = 80 μA); c) volt-ampere characteristic of R2 tunnel diode (X-axis scale: 1 square = 0.1 V, Y-axis scale: 1 square = 2 mA).

The measurement bridge is supplied by the bridge circuit D_1-D_4 through the potentiometer R_1. The measured current passes through the resistor R_3, which has a value of 1 kΩ, and creates a voltage drop which is transmitted to the input of the oscilloscope X-amplifier, while the voltage on the tested device is sent to the input of the Y-amplifier. In order to facilitate observation it is desirable to rotate the cathode-ray tube through 90° or to cross its plates. In testing the forward branch of conventional diodes or in testing tunnel or reference diodes the gain of the Y-amplifier is switched to a value of 1 : 1, thus permitting scale divisions up to 0.1 V/cm.

The reverse characteristics are recorded with a reduction in the sensitivity of the Y-amplifier. For calibration of the sensitivity of the performance-curve tracer on the axes an auxiliary reference diode D_5 of the type D-810 is used with a reference voltage close to 10 V; it is energized by means of the button K_2 (for calibration of the Y-amplifier the tumbler switch is used to change places between the X- and Y-circuits). The two-curve families for triodes are recorded by switching on the base current at 50 cps through the contacts of a type RP-4 polarized relay. Since the collector circuit is actuated at a rate of 100 times per second, two branches are recorded on the screen of the oscilloscope. The switching phase of the relay contacts is regulated by the circuit $R_4 C_1$. The base current of the tested triode is determined by the values of the resistances R_2-R_{15}, which are cut in by means of the switch S_1. For generation of the test-triode base currents a separate rectifier section D_7-D_{10} is used, with stabilization of its output voltage to a value of 20 V by the two diodes D_7 and D_8. Inasmuch as the testing of triodes having a common emitter network essentially requires no change of the base-emitter voltage, every value of the base current is determined by the values of the resistances.

For the recording of the performance curves of n-p-n triodes tumbler switches T_4 and T_3 are provided, which reverse all the polarities of the currents and voltages.

Figures 2a, 2b, and 2c show the volt-ampere characteristics of a triode and tunnel diode, as well as the reverse characteristic of a diode. The instrument is very compact, reliable in long-term service, and has a maximum error of 5-10% in the determination of the parameters, which is fully acceptable for most practical operations.

"POSITION-WEIGHTING" ANALOG-TO-DIGITAL CONVERTERS FOR PULSE AMPLITUDE ANALYSIS

A. N. Volkov, G. I. Zabiyakin, V. G. Tishin, and I. V. Shtranikh

The development of multivariate analysis techniques has necessitated the design of a standard subsystem for the conversion of test pulse amplitudes into discrete digital form suitable for computer processing. The converter must have a channel width error on the order of 10^{-2} for 100- to 500-channel systems and a response on the order of a few microseconds.

The objective of the present article is to explore the possibilities of designing a converter of the indicated type based on the "position-weighting" method. The practicability of this method for multivariate analysis has been argued earlier [1]. A block diagram of the converter is shown in Fig. 1. The network consists of seven similar modules, corresponding to the number of positions of the converter. Each module includes a reference-current generator (I_0), an electronic key, and a control trigger. The current generators are identical in all the modules and are adjusted for a current $I_0 = 20$ mA (in the transistorized version $I_0 = 5$ mA).

The output current pulses are transmitted to an adding ladder consisting of standard resistances R_0 and $2R_0$ of the type often used in computer engineering [2, 3]. It is readily seen that the input resistance of the ladder network on any of the generator sides is $R_{in} = \frac{2}{3}R_0$, and the voltage jump at the point of connection of any of the generators is

$$\Delta U = \frac{2}{3} R_0 I_0 .$$

Since the ladder network forms an attenuator with a stage amplification factor of two, the voltage jump ΔU is attenuated at the output by a factor 2^{n-n_i}, where n_i is the order number of the given stage and n is the number of the highest place. The total output voltage from the network is determined by the state of all the electronic keys, i.e., by the code received by the trigger register.

We now discuss the operation of the converter. In the initial state all the triggers are cleared, except the leading-position trigger. The start pulse, after being transmitted along the delay line sections, takes the triggers sequentially into the "1" state from the highest to the lowest positions. As each trigger is tripped in turn, the output voltage jumps by an amount ΔU_i proportional to the "weight" of the activated position n_i:

$$\Delta U_i = \frac{\Delta U}{2^{(n-n_i)}} .$$

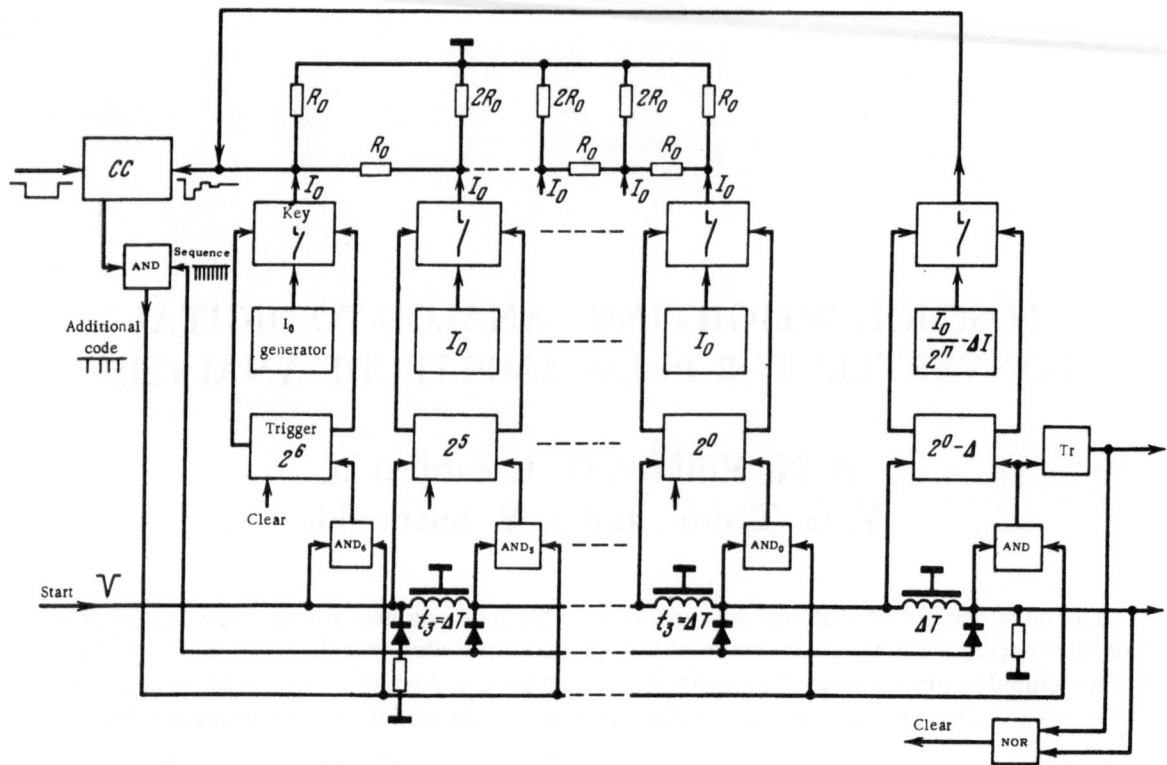

Fig. 1. Block diagram of the position-balance converter.

The total voltage from the ladder output is transmitted to the input of a comparison circuit (CC), where it is compared with the amplitude of the signal to be analyzed. If the instantaneous value of the output voltage from the ladder network at the time of comparison turns out to be smaller than the input signal, the trigger stays in the "1" state, but if it is larger, the trigger goes over the "0" state through the appropriate AND gate. After the last trigger has been activated the digital register has written in it a number in binary code equivalent to the value of the input voltage.

In order to obtain the required accuracy it is necessary for the difference between the instantaneous value U(t) and the adjusted reference value U_{ref} of the voltage at the ladder output, referred to the channel width $\Delta U/2^n$, not to exceed the prescribed error δ in the channel width, i.e.,

$$\frac{U_{ref} - U(t)}{\frac{\Delta U}{2^n}} \leqslant \delta .$$

Inasmuch as the contribution of each position to the value of the output voltage increases with increasing position order number n_i as 2^{n_i}, the accuracy demands of the leading positions grow in the same proportion. The error of the n_ith position must be no greater than $\delta_{n_i} = \delta/2^{n_i}$. It is easily shown that the total resolving time of the converter is determined by the quantity

$$T_{res} = \frac{1}{2} \tau n \ln \left(\frac{2^{n-1}}{\delta^2} \right) ,$$

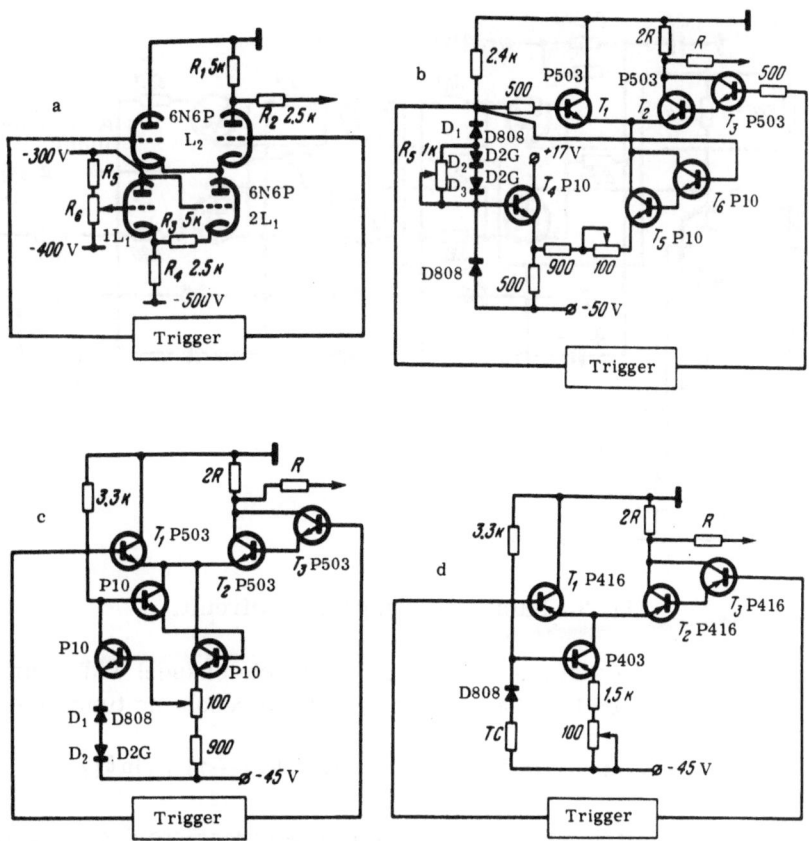

Fig. 2. Circuit diagrams of reference-current generators. a) Vacuum tube balancing network; b) transistorized balancing network; c) stabilizing network with negative feedback; d) stabilizing network with an emitter follower.

and the optimum delay for the ith position is

$$\Delta T_i = \tau \ln \frac{2^{i-1}}{\delta} ,$$

where τ is the time constant of the reference voltages, n is the order number of the highest position, and i is the order number of the given position (i = 1, 2, ..., n).

For our vacuum-tube—transistorized version $\tau = 5 \cdot 10^{-8}$ sec, the maximum delay $\Delta T_6 = 8.1\tau = 0.45$ μsec, and $T_{res} = 1.9$ μsec for $\delta = 10^{-2}$. In the interest of unification and increased accuracy all the delay sections of the converter have been made identical: $\Delta T = 0.5$ μsec, hence $T_{res} = 3$ μsec (for the transistorized model $T_{res} = 6$ μsec).

For greater accuracy and channel-width stability it is advantageous in some cases to use an additional voltage "stage" operating on the Gatti principle [4]. In this case one more module is appended to the converter (see Fig. 1) with a current generator delivering a reference current of such a magnitude that the amplitude of the "step" is somewhat smaller than the minimum channel width $I = I_0/2 - \Delta I$. As a result, the instability of the channel widths is diminished by almost a whole order of magnitude, although narrow "gaps" appear between the channels. The logic network is modified so as to register only those signals whose amplitudes fall within the limits set by the upper and lower levels of the additional "step."

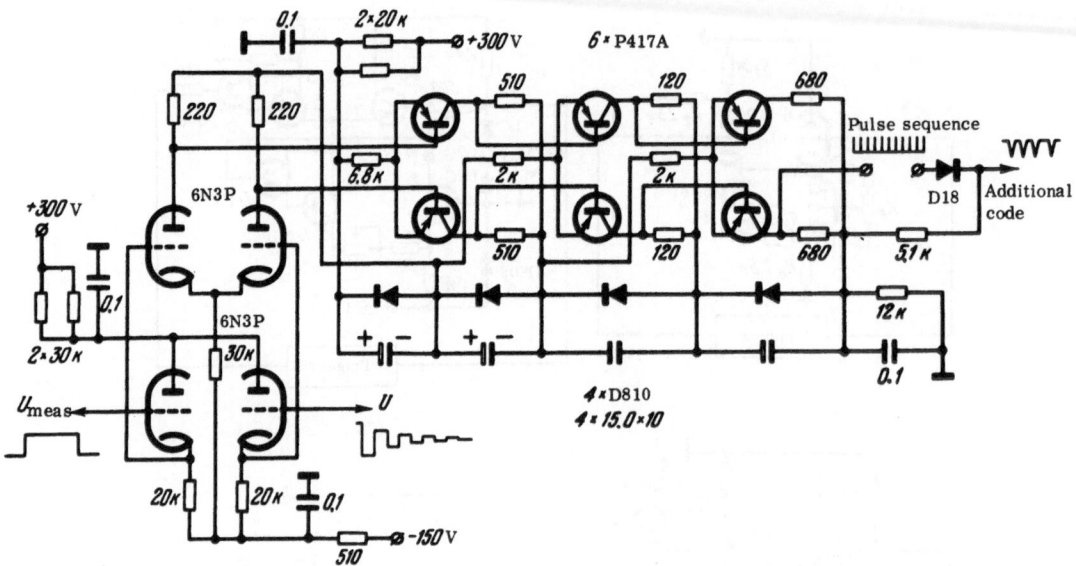

Fig. 3. Balancing comparison circuit.

The accuracy requirements of the converter with stable channel width can now be reduced in the extreme case to ±0.5 times the channel width, and the resolving time is shortened to

$$T'_{res} = \frac{1}{2}\tau_n(n-1)\ln 2 + \tau|\ln\delta| \approx \tau[0.35n(n-1) + |\ln\delta|].$$

For a seven-position converter (n = 6) in the vacuum-tube−transistorized modification

$$T'_{res.} \approx 15\tau < 1 \ \mu sec.$$

In the extreme case, however, the gaps between channels attain values of ±1/4 the channel width. In every specific situation, therefore, it is essential to choose the optimum value of the gaps between channels.

In addition to the errors associated with the finite buildup time of the signal at the output of the resistance ladder, the accuracy of the converter is also directly affected by the stability and precision of adjustment of the position reference−current I_0 generators, the ladder resistances R_0 and $2R_0$, the sensitivity of the coincidence circuit, and the quality of the electronic gates. In our encoders, as mentioned, the total error comes to 1% of the channel width for the vacuum-tube−transistorized version and 2-3% for the all-transistor version.

The reference−current generator (I_0 = 20 mA, Fig. 2a) is made up of type 6N6P bantam dual triodes $1L_1$ and $2L_1$ and includes two cascaded cathode followers L_1 and a balancing electronic gate L_2. Between the follower networks there is a stable potential difference of about 100 V; the same potential is also across the resistance R_3 = 5 kΩ. This connection scheme makes it possible to achieve high stability on the part of the potential difference between the follower cathodes, because it is governed solely by the differential drift of the tubes. In order to ensure stability of the reference current, uniform modes of operation are chosen for the tubes (plate voltage U_p = 100 V, plate current I_p = 20 mA), the resistances R_1, R_2, R_3, R_5, and R_6 are wire-wound or are of the BLP, MGP, or PTMN series with a thermal resistivity coefficient of 10^{-4} to 10^{-5} per °C, and the balancing gate (L_2) is connected to elevated potentials of 20 V, at which the plate current through the blanked tube does not exceed $(2-4)\cdot 10^{-7}$ A.

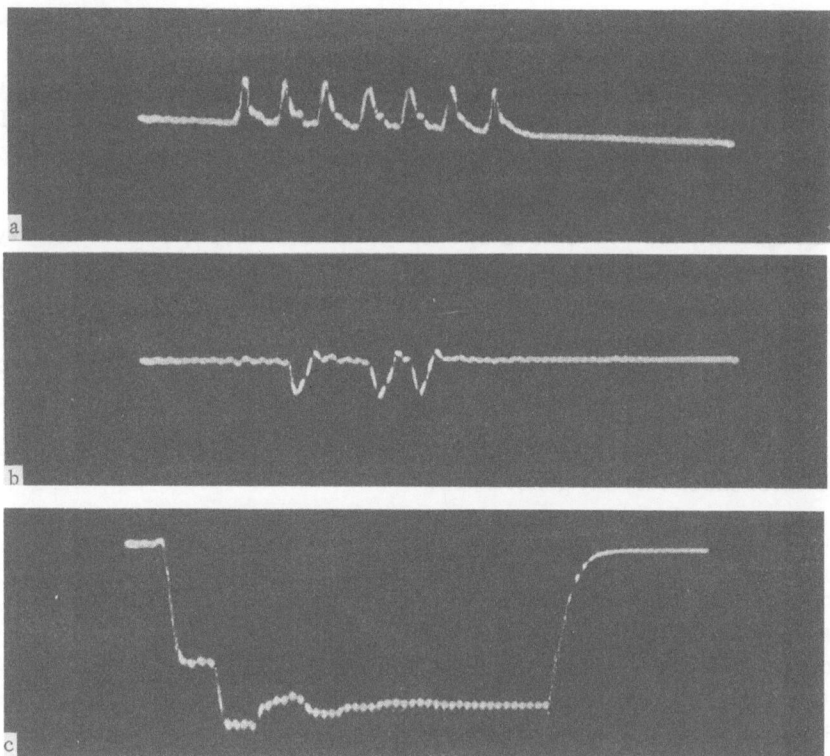

Fig. 4. Oscillograms of the encoder voltages. a) Train of interrogation pulses; b) auxiliary dynamic code pulses to inputs of circuits I_0, I_1, ..., I_6; c) shape of compensating voltage at input to comparison circuit.

To compensate for temperature instability the resistances of the ladder network and the resistance R_3 are chosen at one rating from a single batch. The generator has an instability on the order of 10^{-4} for a supply voltage fluctuation of ±15%. The drift of the reference current does not exceed 3 μA in 8 h of operation. The temperature instability is negligibly small for the vacuum tube version. The analogous transistorized circuit is shown in Fig. 2b.

The temperature instability of the reference current (I_0 = 10 mA) is 0.1 μA/deg. The drift in an 8-h period of operation is about 5 μA. Almost the same parameters are possessed by a reference-current generator constructed on a compensating network (Fig. 2c). The current generator shown in Fig. 2d is assembled on a simpler circuit. Since its parameters are inferior to those of the preceding circuits by almost an order of magnitude, it can only be used in converters utilizing an auxiliary high-stability coding voltage "step." The outstanding feature of all transistorized circuits is the presence of thermal compensation (D_2, D_3, R_5 in the circuit of Fig. 2b, D_2 in Fig. 2c, and TC in Fig. 2d), which is introduced into the circuit of the reference element. All "gates" are constructed on a balancing network. The distinguishing feature of the transistor gates is the inclusion of the additional triode T_3, which increases the output impedance of the current generator and prevents the base current from affecting the collector circuit.

If diodes which each have a low switching time and a reverse current of a fraction of a microampere (D 220, B 14) are tested, the transistor gates may be effectively replaced with diodes

[5]. The comparison circuit (CC) is shown in Fig. 3. The circuit has an input sensitivity better than 1 mV. The stabilizing time of the output voltage is 0.2 μsec. The device is made in the form of a four-stage dc differential amplifier. A 6N3P bantam dual triode tube is used in the first stage to increase the input impedance. The other three stages comprise six P417A transistors. The differential 8-h drift does not exceed 5 mV. Oscillograms of the encoder voltages are shown in Fig. 4.

Experimental tests on the converters showed that the differential nonlinearity for the vacuum-tube—transistorized versions is ±1% of the channel width for 127 channels, and ±2% for the all-transistor version; accordingly, the amplitude coding "dead" time is equal to 3 and 6 μsec, and the drift of the channel widths does not exceed 5 mV for both versions. The maximum input voltage for the vacuum-tube—transistorized converter is 127 V; for the all-transistor version it is 14 V.

In conclusion the authors wish to thank Engineer M. I. Salokhin for assembling and testing the transistorized reference-current generators.

LITERATURE CITED

1. I. V. Shtranikh and A. N. Volkov, Proc. Fifth Sci.-Tech. Conf. Nuclear Radio Electronics, Vol. 2, Part 1, Gosatomizdat, Moscow (1962), p. 10
2. M. L. Klein, H. C. Morgan, and M. H. Aronson, Digital Techniques for Computation and Control, Instruments Publ. Co., Pittsburgh, Pa. (1958).
3. R. K. Richards, Digital Computer Components and Circuits, Van Nostrand, Princeton, N. J. (1957).
4. E. Gatti and F. Piva, Nuovo Cimento, 10:984 (1953).
5. W. D. Rove, IRE Trans. Instr., I-7(1) (1958).

COUNTING CELLS AND CIRCUITS MADE UP OF TUNNEL-DIODE–TRANSISTOR ELEMENTS

V. V. Puzanov and I. V. Shtranikh

Triggers assembled from tunnel-diode–transistor pairs offer the possibility of obtaining acceptably high voltage levels for the control of various logic networks, and they are invested with a good response time and high reliability. As a rule, triggers of this type do not have a counting input.

We propose a trigger network with a counting input and diode-transformer AND circuits (Fig. 1). In this network the transistor can be in the open or closed state, depending on the operating point of the tunnel diode (TD) (Fig. 2). The operating points on the TD performance curve are determined by the volt-ampere characteristic of the TD and load circuit. It can be shown that for the GI305B tunnel diode, given the choice of operating points as shown in Fig. 2, a variation of $E_{sup} = \pm 25\%$ shifts the operating points within the limits of the linear portion of the curve and is acceptable for sharp operation of the counting cell. The TD volt-ampere characteristic fluctuates negligibly with ambient temperature variations, as shown in Fig. 2, hence the operational stability of the cell is not upset between –10 and +40°C.

The network operates as follows (Figs. 1 and 2). Let us suppose that the state of the TD is described by point 2 in Fig. 2. In this case the triode is open, and at point A of Fig. 1 the potential $U_{re_2} \gg U_A \gg U_{re_1}$, so that a positive pulse, transformed from the primary winding W_1 to the secondary W_2, passes through the diode D_1. If the current pulse is greater than ΔI_1 (Fig. 2), it switches the TD into state 1, which closes the transistor. A negative pulse on the winding W_3 cannot pass, and under the given conditions the diode D_3 is still closed, since $U_{re_2} \gg U_A$. After cutoff of the triode current a potential close to U_{re_2} is created at point A, where $U_A \gg U_{re_1}$, so that the next input signal, transformed in W_3 as a negative pulse, passes through the diode D_2 and returns the TD again to the operating point 2, thus reopening the transistor. The collector voltage drops can be used to control the operation of logic networks and for triggering the next cell. In this case the primary of the pulse transformer of the next, completely identical stage is connected to the collector circuit. The integrating capacitor C_1 prevents U_A from varying too appreciably during the pulse period, thus ensuring more sharply defined operation of the network. The response time does not suffer, as $\tau = R_e C_1$ is about 1 μsec. This trigger can be used as the basic element in the construction of an "8 + 2" counting decade.

A schematic diagram of the decade is shown in Fig. 1. The frequency limit of the decade is $f_{lim} = 300$-400 kc, the supply instability is at most $\pm 25\%$ at 100 kc for a temperature variation from –10 to +45°C. It is instructive to analyze the transfer of a pulse in a network of this type comprising a transformer and diode, as it goes from the collector of the preceding cell into the TD of the next one. Inasmuch as the shape of the pulse at the diodes completely determines the delay in the given type of trigger, it is possible to evaluate its response time and stability, as

39

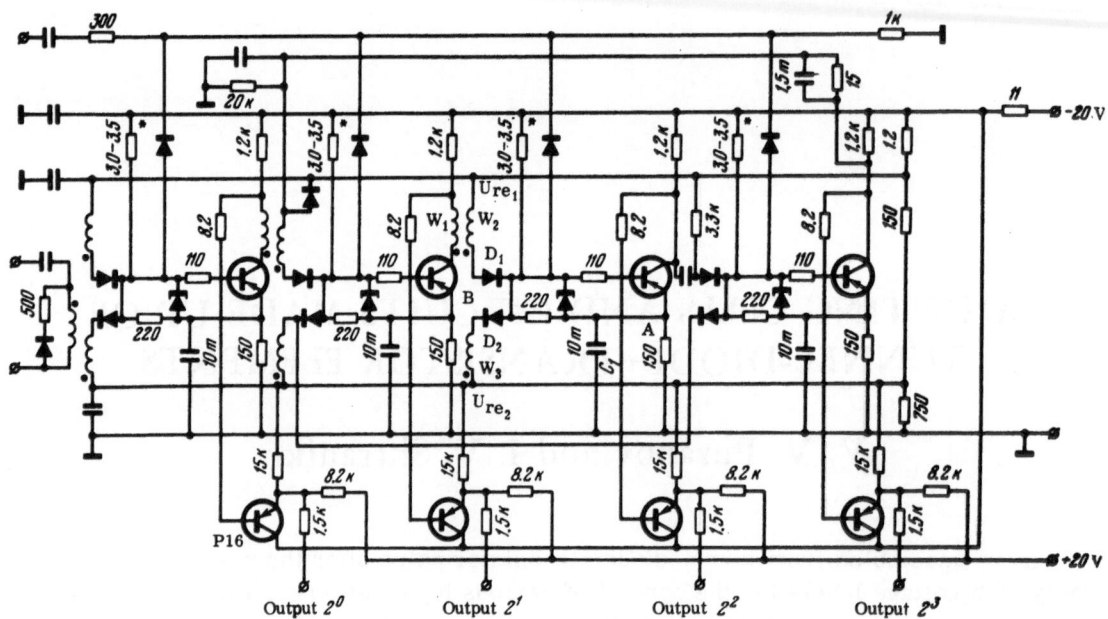

Fig. 1. Schematic of the counting decade. All diodes are of the type D9E; the transformers are $\emptyset$ 10.15 $\times$ 15 $\times$ 30, μ = 1000; the triodes are type MP16B.

well as the influence of variations in the power supply. Either a T-section or II-section can be
used as the equivalent circuit of the triode in analyzing its transient behavior. The fundamental
relations for such a circuit may be found in [1, 2].

We now investigate a T-section (Fig. 3). We examine the shape of the voltage U(t) at point
B (Figs. 1 and 3) under the condition that a voltage jumpe e_0 has been admitted to the base of
the preceding transistor. If we use the customary piecewise-linear approximation for the diode
characteristic, the operation of the circuit may be investigated in three modes.

The first mode, when the voltage on the diode does not exceed U_{lim} (it is composed of the
bias voltage and diode cutoff voltage) and the pulse transformer may be regarded as in the no-
load state. The operator expression for the voltage in the first mode U(0) = 0 may be written
as follows:

$$U(p) = e(p) K_u(p) = \frac{e_0}{p} \frac{\beta'}{(1 + \tau_b - \beta')(R_b + R_e + R_g)} , \tag{1}$$

where $\beta' = \dfrac{\beta_0}{1 + P(\tau_\beta + \tau_{CL})}$, β_0 is the value of β at the middle frequencies, and

$$\tau_{CL} = Z_{CL} C, \quad C = {}'C_C (1 + \beta), \quad \tau_b = \frac{R_e}{R_e + R_b + R_g} ,$$

where Z_{CL} is the collector load impedance.

Equation (1) is readily transformed to the following:

$$U(p) = A \frac{1}{p^2 + 2p\alpha + \omega_0^2} , \tag{2}$$

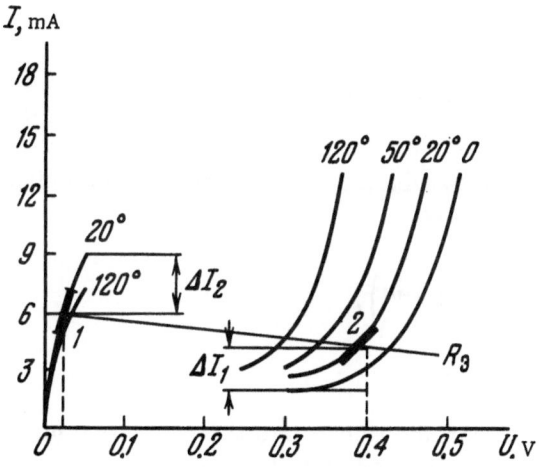

Fig. 2. Volt-ampere characteristics of a tunnel diode for various values of the ambient temperature (R_e is the equivalent resistance of the tunnel diode bias circuit.

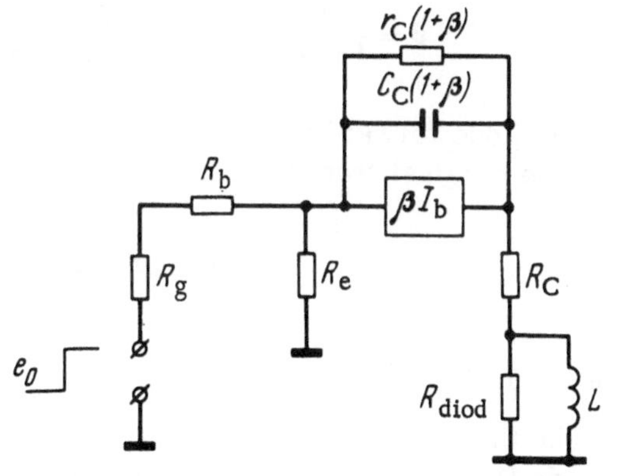

Fig. 3. Schematic of a T-section equivalent to the triode-stage of the trigger.

where

$$A = \frac{e_0\beta}{(R_b + R_e + R_g)(1 + \tau_b\beta_0)C} , \qquad \alpha = \frac{\tau_\beta + \tau_{CL}}{2LC},$$

$$\tau_\beta = \frac{1+\beta}{2\pi f\alpha} , \qquad \omega_0^2 = \frac{1 + \beta_0\tau_b}{LC} .$$

Using operational calculus tables [3], we find the original of the transform (2):

$$U(t) = B(e^{-bt} - e^{-at}),$$ (3)

where

$$B = \frac{e_0\beta_0}{(R_b + R_e + R_g)(1 + \tau\beta_0)\sqrt{\alpha^2 - \omega_0^2}}$$

and a, b are the roots of the equation $p^2 + 2p\alpha + \omega_0^2 = 0$.

When the voltage at point B reaches the value U_{lim} at time t_1, the diode is opened (second mode), and the operator expression for the voltage at this point is written as follows with regard for the initial conditions:

$$U(p) = \frac{R_L^1 I(p)p}{\left(\frac{R_L^1}{L} + p\right)} + \frac{U_{lim}}{\left(\frac{R_L^1}{L} + p\right)} ,$$ (4)

where

$$I(p) = \frac{I_{sat} - t_{t1}}{(1 + p\theta)p} = \frac{I}{(1 + p\theta)p} .$$ (5)

Here I_{t_1} is the value of the triode current at time t_1,

$$\theta = \frac{\tau_\beta + \tau_{CL}}{1 + \beta_0 \tau_b}, \quad \tau_{CL} = C\left(\frac{pLR_L^1}{R_L^1 + pL} + R_C\right), \quad \tau_{CL} = CR_C \quad \text{for } R_L^1 \ll R_C,$$

and R_L^1 is diode forward resistance referred to the primary winding. Substituting (5) into (4) and finding the original from the expression so obtained, we have

$$U(t) = U_{\lim}e^{-t/\tau} + \frac{R_L^1 I}{\theta\left(\frac{1}{\theta} - \frac{1}{\tau}\right)}\left(e^{-t/\tau} + e^{-t/\theta}\right), \tag{6}$$

where $\tau = L/R_L^1$.

The third mode is characterized by the conversion of the triode into the saturation state. The diode D_1 is closed, hence the voltage at point B varies according to the law

$$U(t) = U_{re}\,e^{t/\tau_1}, \tag{7}$$

where $\tau_1 = L/R_C$.

Using Eqs. (3), (6), and (7), we can calculate the shape of the voltage pulse at point B for various values of the transformation ratios n of the transformer. The analytic shape is in excellent agreement with oscilloscope traces. Knowing the voltage pulse, one can find the amplitude of the current pulse that flips the tunnel diode. The calculations yield the following:

for n = 1 t_1 = 0.5 μsec, I_{max} = 8 mA. t_d = 4.5 μsec;
for n = 2 t_1 = 0.3 μsec, I_{max} = 12 mA. t_d = 2.85 μsec;
for n = 3 t_1 = 0.4 μsec, I_{max} = 5.5 mA. t_d = 1.5 μsec.

Here I_{max} is the maximum flipover current that can be developed in the tunnel diode, t_d is the pulse duration at point B, and t_1 is the arrival time lag of the pulse at point B relative to the front of the pulse e_0.

Based on the stipulation of maximum reliability and an acceptable response time on the part of the network, a transformer with a transformation ratio K = 2 was chosen. The response of this type of network is determined by the shape of the pulse transformed into the secondary windings of the transformer. A fast response and high reliability can be obtained by constructing the network on the loop principle with the through transfer of 5 × 2 trigger pulses. The response time of the decade in this case can be made compatible with 700–800 kc.

Considering that each counting trigger of the given type has two "gates," one can vary the feedback in arrays of the triggers in series as a simple means of conversion by 5, 7, etc., or of forming an inverted network.

These networks are distinguished by an excellent response time, reliability, relatively few engineering components, and simplicity of fabrication and assembly. Several decades constructed according to of Fig. 1 are used in the blocks of the FIAN measurement and registration center.

LITERATURE CITED

1. K. É. Érglis and I. P. Stepanenko, Electronic Amplifiers, Fizmatgiz (1961).
2. S. Ya. Shats, Transistors in Pulse Techniques, Sudpromizdat, Moscow (1963).
3. M. M. Aizinov, Transient Processes in Electronic Circuit Elements, Izd. "Morskoi Transport," Moscow (1955).

BUFFER MEMORY DEVICES IN THE MEASUREMENT AND REGISTRATION CENTER

V. V. Puzanov and I. V. Shtranikh

1. Basic Design Considerations

The main memory used in the FIAN measurement and registration center is the magnetic drum of a standard electronic digital computer. This storage system has a number of substantial advantages over other types of storage devices: It permits continuous monitoring of the information store, it is comparatively low in cost and economic to maintain, and it requires a minimum of electronic circuitry. However, the direct use of drum memory for the registration of events randomly distributed in time is highly inefficient due to the large access time (10 msec), i.e., the time to scan and locate the address characterizing a given event.

The speed of these devices is greatly enhanced (50- to 70-fold) by the prestorage of three to five coded events in buffer memory of the flattening (derandomizing) type [1, 2] wherever such events admit almost simultaneous writing and regular sequential readout of information from the contents of this buffer device.

In the terminology of queueing theory [3] such a system (derandomizing device plus magnetic drum) may be represented as a system consisting of one queue with an equal-probability distribution of service times and several coded events arriving at the service point (each event awaiting service comprises one queue element).

A still faster response is attainable by investing the buffer device with, in addition to the derandomizing capability, the property of associativity, which permits those data codes corresponding to the value of the queue number, which is assigned a definite location in the main memory (dynamic type), to be held in the buffer device without disturbing the information pertaining to other events. This kind of device represents a queueing system consisting of several queue elements and several waiting lines.

The structure of the buffer memory device is dictated largely by the arrangement of the data numbers and their digital positions in main memory (as, for example, on the surface of a magnetic drum). The surface can be used in two ways (Fig. 1). In the first case, for the registration of two parameters of a particular event the value of one parameter can be determined by the index number n_i of a large sector of the drum, and the value of the other parameter by an index number m_i inside each sector. A group of like events $z = z(n_i, m_i)$ represents a parallel number written lengthwise on the drum surface. In the second case the value of one parameter n_i is also determined by a large sector, while the value of the second parameter is determined by the track identification number m_i. A group of like events is recorded as a serial code within each track m_i between the limits of the sectors n_i. Intermediate variations of storage are also possible.

43

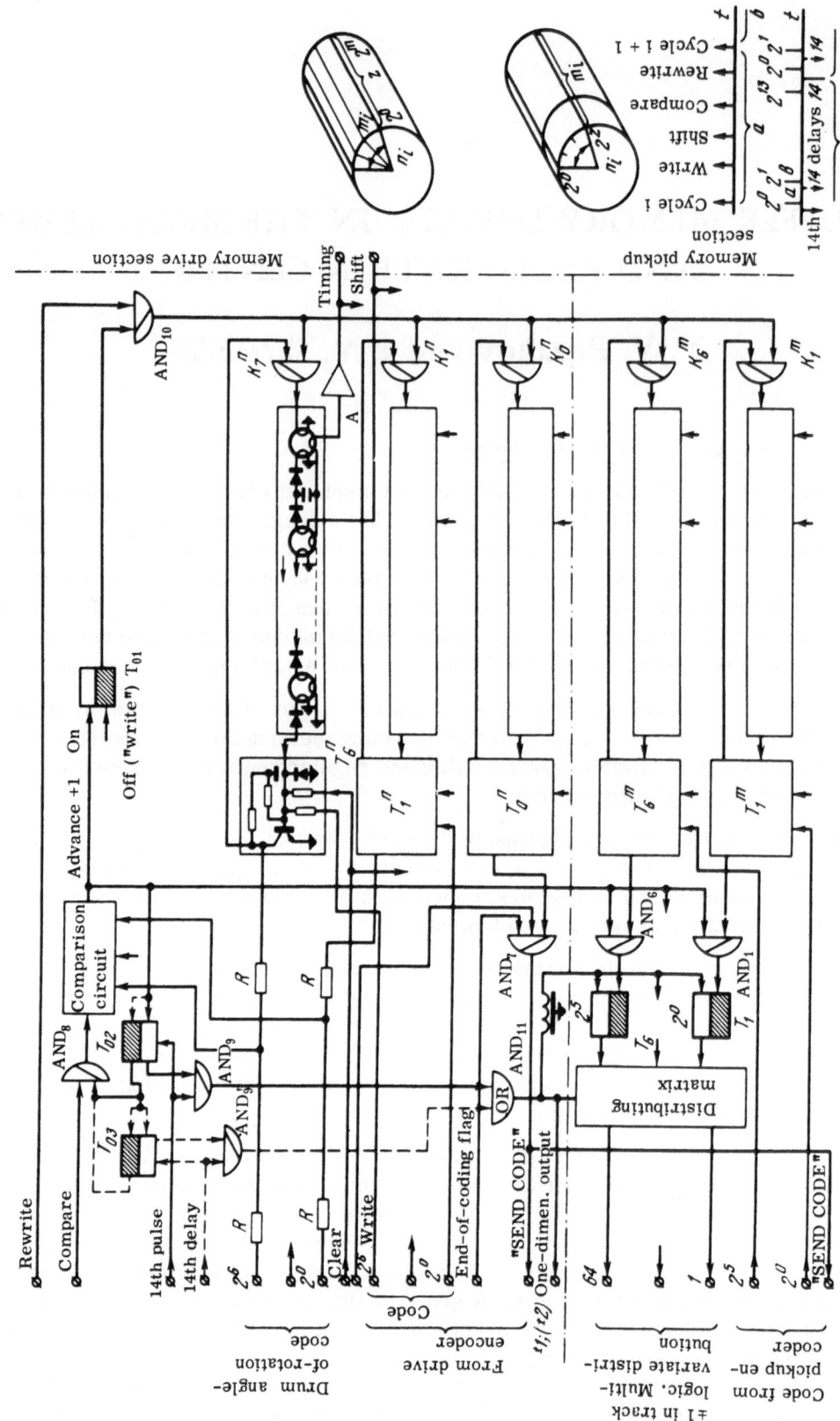

Fig. 1. Diagram showing the action of the buffer memory blocks on the shift registers for a multivariate analyzer.

The advantage of the first method of allocation is parallel acquisition of the number z, a form that is convenient for readout. Its shortcomings include: a) low reliability, since the breakdown of one position of z distorts the entire set; b) the need for the buffer memory system to have comparison with the value of the drum angle-of-rotation code for the entire set of numbers; c) fewer possible combinations of z.

In our actual system we decided on the second alternative (see Fig. 1). The MRC permits several multidimensional and one-dimensional measurements to be performed concurrently, where the relative number of measurements of a particular type can be varied from time to time. Consequently, the buffer memory devices must have multidimensional and one-dimensional branches, which are easily adjusted to meet the demands of the experimental problem.

2. Structure of the MRC Buffer Devices

As mentioned, the fastest response is obtained from buffer memory units having the property of associativity. This property can be realized in two ways: a) by using memory systems that permit the rewriting of data codes that do not match the current address code at a given time; b) by using special nondestructive readout elements executing the mismatch logic operation $A\bar{B} - \bar{A}B$ (or match $AB + \bar{A}\bar{B}$), where A is the stored information and B is the reference information.

The first type of memory is realized when a shift register with a code rewrite facility is used as the accumulator. In this case every data code received from the encoder is cycled continuously in the register until it coincides with the instantaneous address code; in the latter event it is taken out of the register into main memory. This principle has made it possible to build reliable and inexpensive buffer memory units for the MRC, which are quickly adjusted for multivariate or univariate analysis problems. Simple commutation networks permit several buffer units to be operated in parallel, thus increasing the volume of stored information and, therefore, diminishing the counting losses in a particular measurement system. This type of memory is recommended when the registration rate is not too high (from 200 to 800 pulses/sec). For a high rate of counting the access time to a free address for writing of the next data code from the encoder becomes considerable, causing a disproportionate increase in the volume of electronics and a further undesirable growth in the physical size of the memory.

The second type of memory device, which uses nondestructive readout elements, is free of the aforementioned shortcomings. It is important to note that the problem is most completely solved by the use of nondestructive readout elements that execute the mismatch logic operation. Some of the elements already in use include ordinary toroidal ferrite cores, for which readout is effected by short current pulses (50 nsec) [4]; elements developed by the electrical simulation laboratory of the All-Union Institute of Scientific and Technical Information (VINITI) of the Academy of Sciences of the USSR [5] utilizing a "figure-8" construction; and double-layer magnetic films [6]. Analysis has shown that of all these elements the best signal-to-noise ratio in combination with a good response time is possessed by figure-8 ferrite elements. For this reason a multidimensional associative memory using 25 14-position numbers and having an input capability to handle 2000 random pulses per second at 1% counting error has been developed for the MRC.

3. Buffer System for Multivariate Analysis

on Shift Registers

We now examine the case of analysis in which the registered event is determined by two coded parameters having $2^m - 1$ discrete values for one parameter and $2^n - 1$ discrete values for the other. Then the parameter characterized by a position on the periphery of the drum (see Fig. 1) is the drive parameter, and the parameter characterized by the track identification number is the sense parameter. This calls for two converters, a drive analog-to-code and a sense analog-to-code. The coding time of the drive encoder in this case must be greater than that of the sense encoder, hence simpler logic networks are permitted.

The actual memory system contains seven main and one auxiliary register in the drive network and six registers in the sense network (spectrum on 128×64 channels). The auxiliary register is used to transmit the descriptor. All positions of any number are entered into the registers in parallel, and for "1" values the output is switched to the triggers T_0, T_1^n-T_7^n, T_1^m-T_7^m. The entire system is controlled by a sequence of selector pulses (see Fig. 1), which are spaced 2-3 μsec apart for an internal cycle of 50 kc.

If the encoding devices generate codes, then the "end-of-coding" flag voltage is transmitted from the drive encoder to the AND_7 unit. After the "timing" pulse, if there is no number in the given row of ferrites (no code-present descriptor), the trigger T_0^n does not energize, and a "write" gate pulse is transmitted through the AND_7 unit, generating a command to the drive and sense encoders to SEND CODE. At this time a parallel binary pulse code arrives from the encoders at the same triggers T_0^n-T_7^n and T_1^m-T_6^m, and the encoders are cleared. The potential code thus formed at the triggers T_0^n-T_7^n of the memory drive section are compared in the comparison circuit with the drum angle-of-rotation code, advanced by one, corresponding to the given time. If the code coincides with the angle-of-rotation code, at the instant of arrival of the "comparison gate" pulse a preliminary "+1" signal is generated. The preliminary signal, after passing through the transmission networks AND_1-AND_6, transmit the potential code from the memory sense block to the input triggers T_1-T_6 of the track-distributing matrix, which is a standard diode decoder with $2^6 - 1$ outputs. Simultaneously this pulse trips the trigger T_0, which acts in turn to inhibit the rewrite cycle, and then the ensuing "clear" pulse takes the triggers T_0^n-T_7^n and T_1^m-T_6^m into the initial state, thus extracting the number from the buffer memory. If the codes for the given angle of rotation of the drum and the preceding number do not match, the rewrite circuit, controlled by the trigger T_{01}, sends a pulse to gate circuits K_0^n-K_7^n and K_1^m-K_6^m, which are controlled in turn by the respective triggers T_0^n-T_7^n and T_1^m-T_6^m; also the preceding code or the code taken from the last ferrite cores of the shift registers are written in the input cores of the registers, and the triggers T_0^n-T_7^n and T_1^m-T_8^m are switched after a certain period of time into the initial state by a "clear" gate pulse. The number written in the cores of the shift register are continuously shifted by the timing and shift pulses. During the interrupt time of the drum within the limits of one channel (about 160 μsec) there is time to scan seven words in the buffer memory in the 14 arithmetic positions in each channel. Then the gate pulse for the 14th position functions through the distributing matrix to generate at one of the outputs of the latter (in correspondence with the code written in the triggers T_1-T_6) a "+1" signal, which is then sent to the totalizer logic network of the drum for a given track before the occurrence there of the position 2°.

Inasmuch as several identical numbers can occur in memory, with the occurrence of the preliminary "+1" signal in the network of the standard version the trigger T_{02} is tripped, thus inhibiting the comparison gate pulse, whereupon no number can be extracted from buffer memory, and only the rewrite operation remains. The nearest 14th-position gate pulse returns T_{02} to the initial state.

In order to speed up the response time (for the registration of spectra having spikes, in which case the probability of identical codes occurring is greater), it is possible to include in the system addition by ±1 and ±2. For this purpose a 14th pulse, delayed by one arithmetic position, is used; it is transmitted to the logic networks of the adders to bring in the second position at once for processing. For this, as indicated by the dashed line in Fig. 1, a trigger T_{03} with counting inputs is furnished. The trigger T_{02} is also converted to the counting mode.

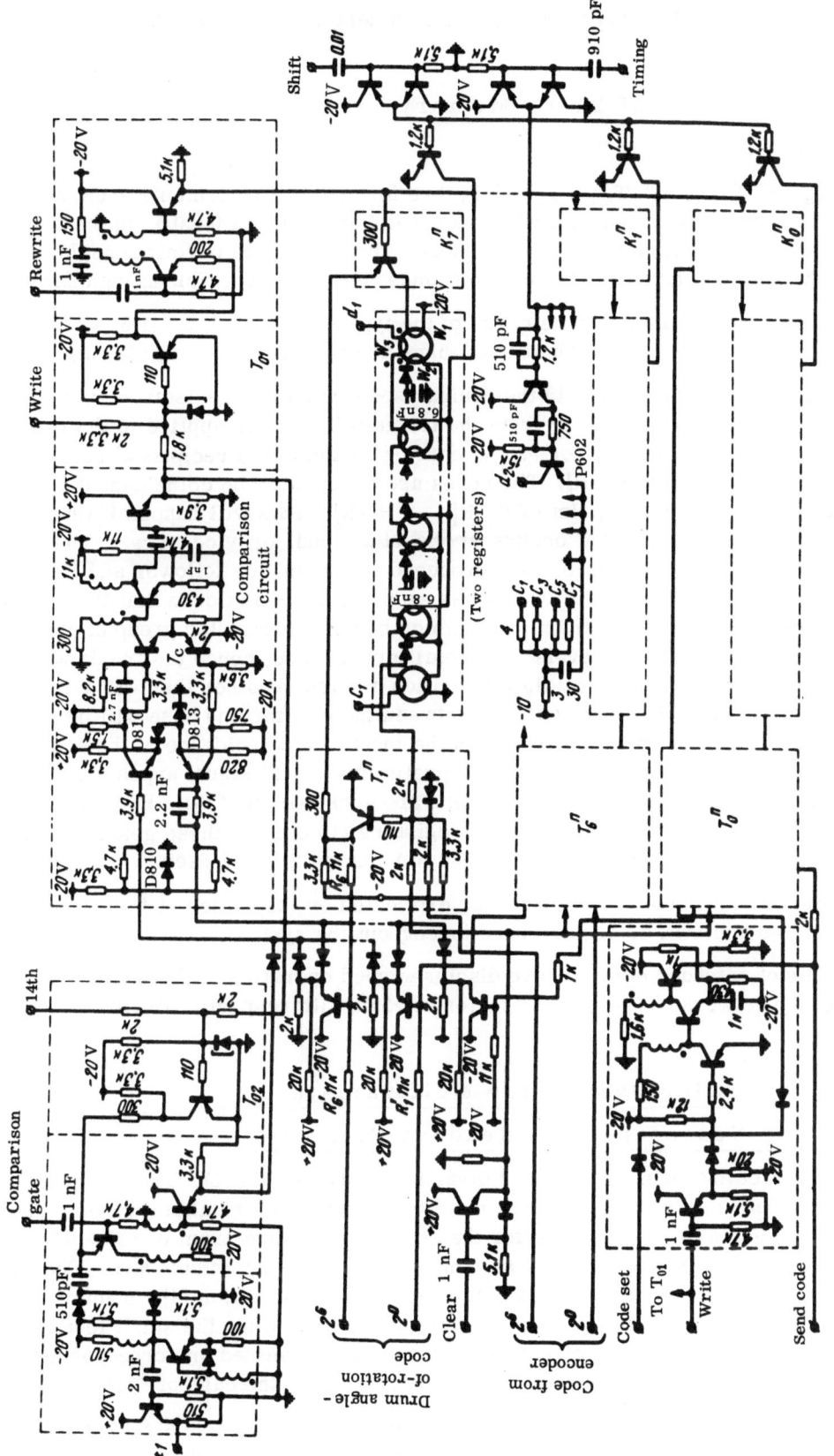

Fig. 2.   Schematic diagram of the drive section of the shift-register buffer memory device.

It is readily seen that the comparison gate pulse is inhibited only when the comparison circuit puts out a preliminary "+1" signal twice and the 14th delayed signal is admitted to the input of the adder.

4. Brief Description of Drive Memory

Network (Fig. 2)

The triggers T_0^n-T_7^n, T_1^m-T_6^m, T_{01}, and T_{02} are formed of two elements each, a 31301G tunnel diode and a type P16 transistor, which generate potentials of 0-20 V on the triode collector, depending on the polarity of the input pulse. These triggers permit considerable simplification of the network, a reasonably fast response time, and superior gating characteristics.

The shift registers utilize ferrite cores of the type K-272 4×2, 5×1.5, which have a narrow loop ($H_c = 0.21$ A/cm). The number of turns are $W_1 = 20$, $W_2 = 40$, $W_3 = 3$.

The basic comparing elements in the comparison circuit are resistance pairs R_1'-R_1, . . . , R_6-R_6'. The voltage levels (0 or -20 V) from the triggers T_1-T_6 are applied to the resistances R_1-R_6, while the potentials (again, 0 or -20 V) corresponding to the reciprocal code of the drum angle of rotation are applied to R_1'-R_6'. The codes are assumed to be coincident if a potential of 10 V is created at the common point of R_1-R_1' . . . R_6-R_6'. This will happen if when 0 occurs at one end of these "weights", -20 V occurs at the other end, or vice versa. The potential of the common point of the resistances is reproduced by an emitter follower and transmitted to two assemblies per digital position of two diodes each. It is seen at once that if the codes coincide, a potential of -10 V will appear on the common bus of a particular group of diodes. In this case T_{com} will have the same levels at the emitter and base, hence the triode will be closed. In any other case, even if for just one digital position a potential other than -10 V is obtained at the common point, the ratio of the voltages between the emitter and base will be such as to open the triode T_{com}. To sharpen the operation the instant of coincidence of the codes is strobed. Consequently, if the codes coincide, a preliminary "+1" signal appears at the output of the comparison circuit. The timing generators are constructed of type P602AI triodes with cooling radiators, one triode serving two shift registers (10 cores); these generators permit the formation of a current on the order of 70-800 mA lasting 2.5 to 3 μsec.

The operation of all other circuits can be explicated by inspection of Figs. 1 and 2. Type P16 and P10 triodes and D9K diodes are used throughout the system.

The results of extended service have disclosed good operation of the network over a long period of time; the networks have a safety factor of $\pm 15\%$ with respect to fluctuation of the supply potentials.

5. Prestorage Buffer System for Multivariate

Analysis Using Nondestructive Readout Elements

As in the preceding case, we consider bivariate analysis (Fig. 3). The system is designed to handle a 7-position code with an auxiliary position for writing a number-present descriptor in the drive branch and a 6-position code for the sense branch (spectrum of 128×64 channels). The memory is capable of simultaneously storing up to 25 14-position numbers. The base of the system is a matrix of ferrite elements with a figure-8 construction [4]. If information in zero-one form is written on the large loop of the figure-8, then when the small loop is scanned the value of the output signal on the signal winding can be determined on correspondence with "0" or "1" as "no" in the case of "00" and "11" and as "yes" in the case of "01" and "10."

In order to improve the signal-to-noise ratio and to create a more stable operating mode for the destructive readout generators, here a system with linear sampling of the number from the record by a (2/3 - 1/3) system with two cores per position is used.

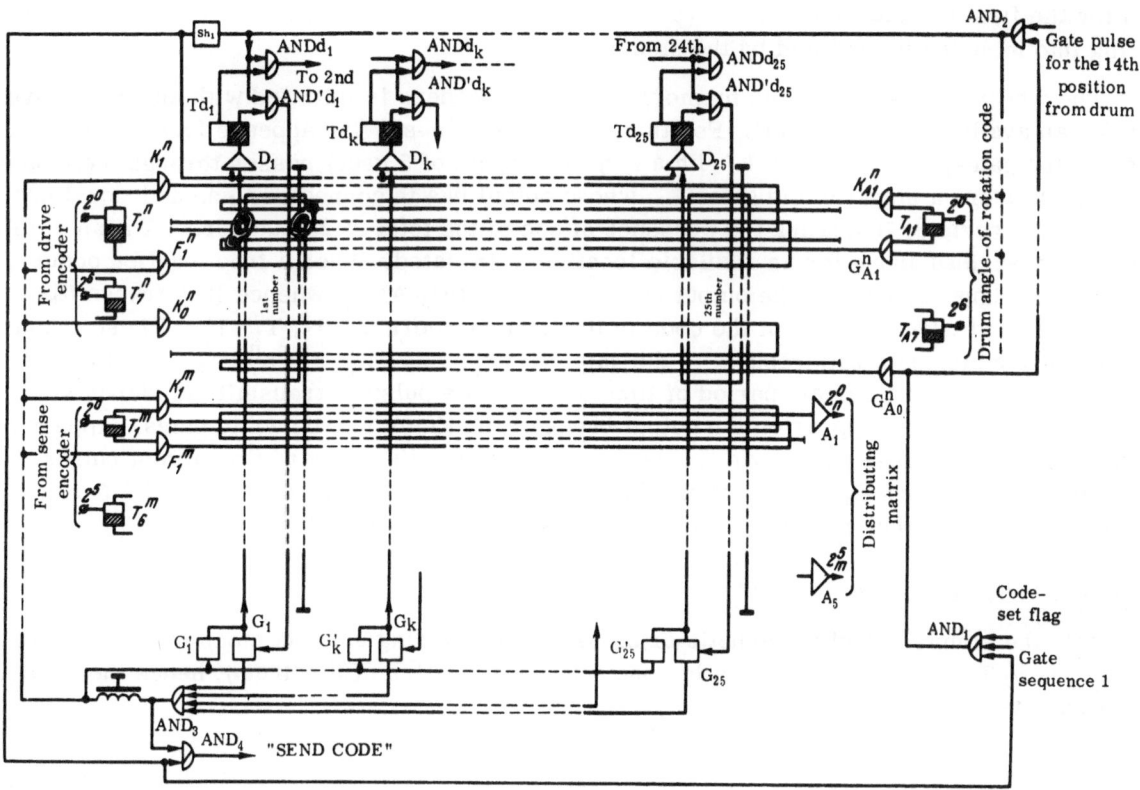

Fig. 3. Block diagram of the prestorage buffer system using nondestructive readout elements.

A simplified block diagram of the memory system is shown in Fig. 3. The drum angle-of-rotation code, advanced by one, is admitted to the triggers T_{A1}-T_{A7}. The trigger for each position is the controlling member with respect to a gate pair of the type K_A^n-F_A^n. At the time of arrival of the 14th-position gate pulse from the drum, on the output buses of gates K_A^n or F_A^n there appear scanning pulses ($t_i = 0.15$ μsec, $I_{re} = 1$ A) to be transmitted to the matrix. If a code coincident with the given drum angle-of-rotation code is stored in the matrix, no signal other than a noise signal is generated in the signal winding for that number. In the signal windings for numbers differing by at least one position 100–150-nsec pulses with an amplitude of 30 mV appear. If the number of noncoincident positions is large, the signal amplitude is correspondingly greater. Amplified and gated in the detectors D_1-D_{25}, these pulses trip triggers $Td_1 = Td_{25}$. In this way numbers that do not coincide with the address code correspond to the activated triggers of the whole set Td_1-Td_{25}, and vice versa; wherever the codes coincide the triggers are not tripped. Then a serial scan of the first coincident number takes place. For this a special scanning pulse generated by the network Sh_1 transmits a pulse to the modules $ANDd_1$-$ANDd_1'$; if a trigger happens to be tripped (meaning that the given number did not match the address code), at the output of $ANDd_1$ there appears a pulse, which is transmitted to the next modules $ANDd_2$-$ANDd_2'$.

It is seen at once that the initial scanning pulse moves serially from one number to the next until it arrives at an untripped trigger (meaning that the given number has coincided with the address code). In this case the signal comes to a stop at that point and triggers the destructive readout generator G_k through the module $ANDd_k$, whereupon signals are formed at the outputs of the amplifiers A_1-A_6 to determine the sense encoder code, which at some time has been written into this memory cell. Subsequently this code is transmitted to an ordinary diode with $2^6 - 1$ outputs, which is a drum track distributor. The total time to locate the number, scan for the first coincidence, and extract it from memory is about 3 μsec. During this period writing into memory is inhibited by the module AND_1.

The writing of a number into memory from the encoders is initiated with an associative scan for an available memory cell. For this, when the code-set flag appears from the drive encoder, the gate-pulse sequence 1 (with a sequence cycle of 5 μsec) passes through the module AND_1 and delivers a signal to the gate F_{A0}^n, whose output bus is connected to an additional position for the descriptor indicating the presence of one of the total of 25 numbers. A scan similar to the one described above for an available location is initiated. When a free location occurs, a SEND CODE signal appears at the output of the module AND_3–AND_4, whence it is transmitted to the encoders and takes from them the code to the input triggers T_1^n-T_7^n, T_1^m-T_6^m. These triggers are controlling elements with respect to the gates K_1^n-F_1^n; . . . ; K_7^n-F_7^n; K_1^m-F_1^m . . . ; K_7 -F_7 . Therefore, when after a certain period of time a write gate pulse is transmitted, 300-mA 1.5-μsec current pulses appear on the output buses of the K- or F-type write gates. Appearing simultaneously with these pulses via the destructive readout lines are write current pulses developed by generators G_1-G_{25}, which produce 600-mA currents with a duration of 1.5 μsec. The operation of all other sections of the system is self-explanatory from Fig. 3.

6. Counting Errors in Buffer Memory

With the simple use of a magnetic drum as the registering unit the mean memory access time T_{acc} is $\frac{1}{2}$ T (where T is the time for one complete rotation of the drum), hence the relative loss of registering events is

$$L = \frac{N_{in} T_{acc}}{1 + N_{in} T_{acc}},\tag{1}$$

where N_{in} is the input rate of events to the system. if T_{acc} is on the order of 10 msec, then for a counting rate of just 1 pulse/sec already 1% losses are possible. In the language of queueing theory such a system is equivalent to a one-line queueing system (n = 1). In our case, however, the buffer devices can store five distinct numbers in the case of a memory built of shift registers, and 25 numbers in the case of a memory using nondestructive-readout cores and the principle of associative interrogation in the scanning of these numbers. In addition, the encoder contains one register, which stores a coded number until a location buffer memory is available. From the point of view of queueing theory, therefore, these devices represent a system comprising one service point and several lines. In the general case the counting losses in such systems where several service points are involved depend on the law governing the arrival of these events at the system input and the service time distribution. Given a Poisson input flow and a positive service time distribution g(t) = $\mu e^{-\mu t}$ (where μ is the rate of service per request), we have

$$\mu = \frac{1}{M(T_{serv})},$$

where $M(T_{serv})$ is the mathematical expectation value of the service time. For the relative counting error we have [7]

$$L_{n+m} = \frac{\frac{\alpha^n}{n!}\left(\frac{\alpha}{n}\right)^m}{\sum_{k=0}^{n}\frac{\alpha^k}{k!} + \frac{\alpha^n}{n!}\sum_{s=1}^{m}\left(\frac{\alpha}{n}\right)^s},\tag{2}$$

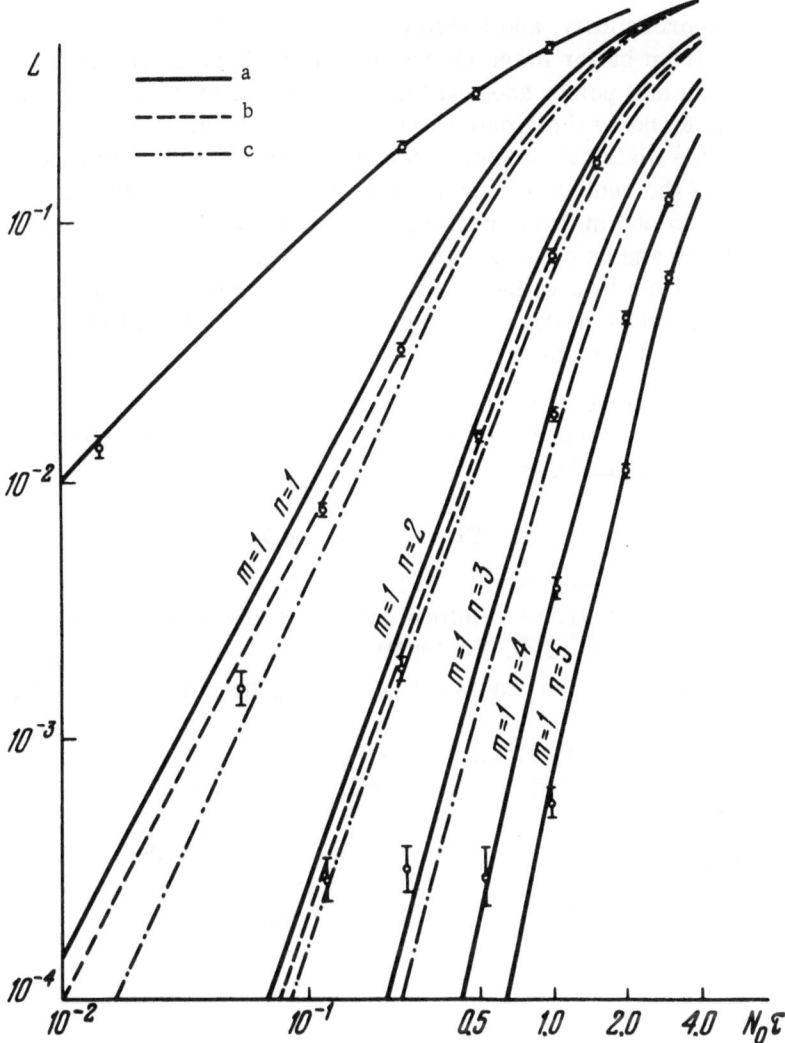

Fig. 4. Counting error in buffer memory. a) Exponential service distribution; b) equal-probability service distribution (from 0 to T); c) constant service distribution.

where n is the number of lines and m is the number of service points. Curves depicting the case m = 1 and n = 1, 2, 3, 4, 5 are shown in Fig. 4.

For other service time distributions, such as T_{serv} = const or for T_{serv} assuming equally probable values from 0 to T, Eq. (2) cannot be used in the strictest sense to compute the error. Only in the special case when the system comprises single lines, according to the Sevast'yanov-Khinchin theorem [8], and the input flow has a Poisson behavior does the service time distribution law become insignificant. Unfortunately, queueing theory does not offer a solution in general form to the problem of the errors in queueing systems involving multiple service points and multiple waiting lines. The most auspicious approach appears for the time being to use statistical modeling (Monte Carlo) techniques for the analysis of such systems.

The given queueing system has been simulated on a computer. The results of computations for the case of one service point and 1, 2, 3, 4, and 5 lines are shown in Fig. 4 for various service distribution laws. It is apparent from the graphs that as the number of lines increases relative to the number of service points (one service point), the influence of the service distribution diminishes, and, beginning with just four lines and one service point, all three curves (for exponential, equal-probability, and constant distributions) merge into a single curve. The extraction of numbers from buffer memory onto magnetic drum is realized by an equal-probability law. The experimental points adopted for the system analyzed above provided a good fit to the analytical points found by the Monte Carlo method. Examination of these curves for an equal-probability service distribution also discloses, using Eq. (2) with a nonexponential service distribution, that if the number of service points is much smaller than the number of waiting lines, it is possible to obtain fairly precise values of the relative errors in queueing systems. All measurements and calculations were carried out for the case of a uniform input spectrum and a Poisson time flow of incoming events. Referring to Fig. 4, we readily notice that for such an input spectrum a 1% counting error occurs for a load rate $N_{in} = 200$ pulses/sec in the case of a shift-register memory.

For a memory constructed of ferrite elements with nondestructive readout, calculations based on Eq. (2) show that the input rate for 1% counting error should be about 2000 pulses/sec for a memory capacity of 25 numbers.

LITERATURE CITED

1.   M. I. Lapin, Avtomat. i Telemekhan., 23:321 (1962).
2.   V. V. Puzanov and I. V. Shtranikh, Inform. Byull. SNIIP, Komitet po Ispol'zovaniyu Atomnoi Énergii SSSR, No. 16, p. 36 (1966).
3.   V. V. Puzanov, I. V. Shtranikh, and A. T. Matachun, Pribory i Tekh. Éksperim., No. 3, p. 82 (1966).
4.   J. R. Kiseda, IBM J. Res. Develop., 4:106 (1961).
5.   E. I. Il'yashenko and V. F. Rudakov, Components, Modules, and Devices of Information Processors, VINITI, Moscow (1964).
6.   L. J. Oakland and T. D. Rossing, J. Appl. Phys., 30:545 (1959).
7.   E. S. Venttsel', Probability Theory, Fizmatgiz (1960).
8.   B. A. Sevast'yanov, Teor. Veroyat. i ee Prim., 2:106 (1957).

INFORMATION STORAGE DEVICES IN THE SYSTEM OF THE MEASUREMENT AND REGISTRATION CENTER

A. M. Klabukov and I. V. Shtranikh

In the measurement and registration center [1] magnetic drum storage is used for the large-capacity main memory. The storage device in this case has 88 tracks and is equipped with 88 heads, which are distributed over opposite sides of the drum. There is one erase head comprising three independent sections. The circumference of the drum is about 1 m, and its speed of rotation is about 25 rps. The recording density can be as high as 3 pulses/mm. The track width is approximately 2 mm. A sequence of 3072 pulses at a frequency of about 80 kc is provided for operation of the computer. We have realized a 3584-pulse sequence, which permits the allocation of 14 digital positions per channel on each track of 256 channels for analyzer applications. The recording of signals on the drum surface is accomplished at two levels with interrecord gaps or return to zero at a timing frequency of approximately 90 kc [2].

The surface of the drum is most fully utilized (in the case of nonerasing operation) by operation on the one-to-zero and zero-to-one rewrite principle, but by means of a single head. Inasmuch as the signal to be read is the derivative of the drum surface remanent magnetization lengthwise along the track, it occurs earlier than the time (location) corresponding to the exact

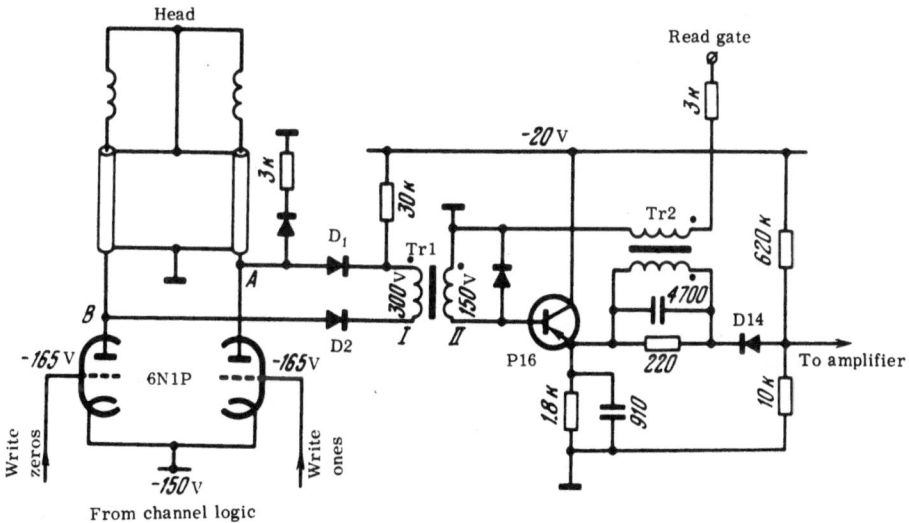

Fig. 1. Amplifier network for operation with a single read–write head and no erase head.

phase position of the head gap at the write time relative to read time. At this instant the read signal is zero, then it changes sign.

For a drum with a gap of 30-40 μ between the head and its support and a write pulse duration of 0.8-1 μsec the delay of the maximum level of the read signal is 2-4 μsec, which is adequate for the execution of arithmetic operations in the logic section. This method should also be desirable in its potential elimination of the erase head. However, implementation of the method is frought with several problems.

1) It is difficult to build a nonoverloading read amplifier with an overload factor up to 10^3.

An amplifier network designed to meet these requirements is illustrated in Fig. 1. Its main feature is the inclusion of diodes D_1-D_2, one of which is closed during the writing of "zero", the other during the writing of "one," thus inhibiting direct transmission of the write signals to the amplifier input. Another attribute of the circuit is the incorporation of a gating signal to the amplifier circuits; this improves the phase relations. The given amplifier has a "dead" time of less than 10 μsec.

2) The maximum delay time allowable in the amplifier is 1 μsec.

3) The rewrite principle makes for a poor one-to-zero ratio, reducing it to 3-5, as opposed to a ratio of 20-40 for operation with an erase head.

4) The most undesirable effect occurring in writing is magnetic saturation of the head core, as this lowers the permeability of the latter, thus deteriorating the coupling of the head with the support magnet and tending to diminish the next read signal.

Figure 2 shows as an example the dependence of the decline in amplitude U_r of a read signal with augmented magnetization of the core by a constant current I. The signal is not erased in this example. As is apparent from Fig. 2, not very many ampere-turns are required to completely suppress the read signal.

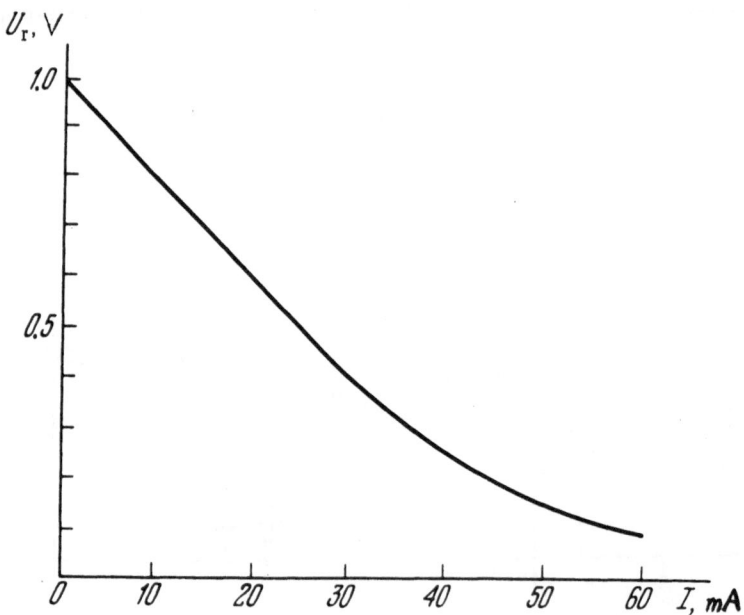

Fig. 2. Read signal amplitude U_r versus the current I for augmented magnetization of the head core. Gap = 30 μ.

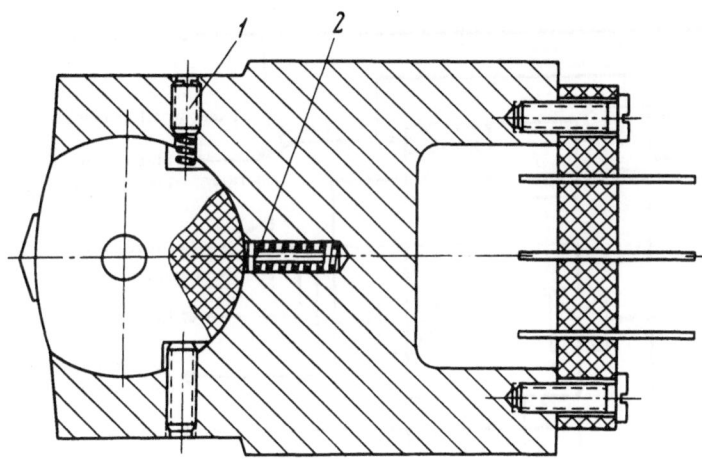

Fig. 3. The modified magnetic head.

Regeneration of the magnetic flux in the core takes more than 20 μsec, creating a corresponding emf that is fairly easily compensated. However, the dead time of the write cycle is then more than 20 μsec, hence the storage density indicated above (3584 pulses/track) cannot be realized, as it only allows for a write cycle of at most 11.4 μsec. The method could be used under the given conditions with a reduced storage density, but the small margin left for the other parameters would decrease the reliability of the system, therefore it is preferable to work with two independent read and write heads.

As indicated above, the heads, of standard factory construction, are distributed over both sides of the drum on alternating tracks. Consequently, unless one resorts to two independent erase heads and a special chopped magnetic field configuration, the most sensible approach is to arrange the heads on diametrically opposite points of the drum. This arrangement has the following attributes, over and above the possibility of advancing the read signal significantly and the elimination of effects of the write signals on the read circuits: With a strict 180° separation of the heads (without regard for the lead of the read head) the system can function without an erase field. The memory access time is halved, amounting to 1/50 sec, i.e., the number of channels addressed per unit time remains unchanged.

The permissible error in the relative mounting of the pair of read and write heads is ±1 μsec (with respect to the signals), which corresponds to less than ±0.01° in angle, or about ±27 μ on the drum circumference. This mounting precision is achieved by final adjustment of the circumferential position of one of the heads of the pair. For this the heads are slightly modified (Fig. 3). One of the set screws is replaced by a spring-loaded detent 1. Also, in order to maintain a constant gap, an expansion spring 2 is inserted to take up the play in the yoke of the holder. In this system final adjustment is possible, without appreciable signal reduction, between the limits of ±(15-20) μsec. For the preliminary mounting of the heads a single pulse is written simultaneously on both heads of the pair. The spacing of the heads is quickly ascertained from the relative time separation of the read pulses. The read head in the proposed system must lead the write head by 7-9 μsec.

At first the proposed write system [3] was intended for the parallel recording of a number by the use of many tracks at one time. In the final version writing is executed serially in 14-position binary code in each of 128 channels for each of 80 tracks (8 tracks serving as auxiliaries). This kind of system is easier to operate and service. It permits step-by-step incorporation of the various branches of the center as they become operational and ensures the capability of flexible allocation of the system capacity.

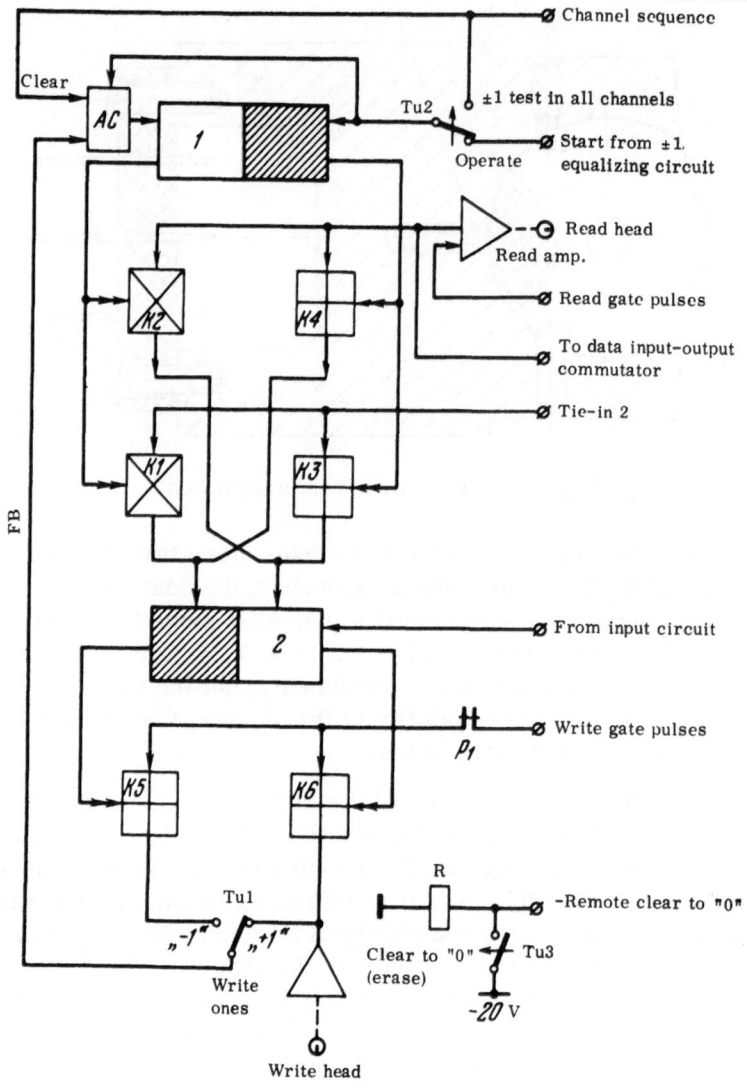

Fig. 4. Block diagram of the channel information storage
block.

For synchronization and control of the operation of all essential devices of the MRC three
main sequences of pulses are written on the auxiliary tracks of the drum: drum half-turn
pulses, channel pulses, and arithmetic pulses, where the second of these are rigorously matched
in phase with every fourteenth pulse of the third sequence. Writing of the indicated pulses on
each of the auxiliary head pairs is performed concurrently. Writing takes place during one
half-turn of the drum by means of a special simultaneously-operating device. The latter comprises an oscillator with a regulatable frequency on the order of 90 kc, gates that are closed
by the drum half-turn pulse (with a 20-μsec delay) and opened by the last write pulse (128th) of
the channel sequence, two series-connected 14-way and 128-way dividers, and an inhibiting
device to prevent retriggering.

During adjustment the frequency of the generator is set in such a way that the end of the
pulse sequence being written and the next half-turn pulse are separated by a time interval on
the order of 50–80 μsec (as read on an oscilloscope). In order for every 14th pulse of the arithmetic sequence to precede its channel pulse, during the writing of these sequences the drum

rotates in the direction opposite the normal operating direction. In their operating position corresponding pairs of heads are run in parallel, so as to average out the phases and amplitudes of the pulse sequences being read in the event of any slight misalignment of the heads at 180° separation. Together with the phasing half-turn pulse the channel sequence is transmitted regularly to the central-station channel identification number counter, whose instantaneous code is delivered to the code comparison circuits of the derandomizing devices and to the common subsystem for various types of data output. The counter is cleared by the drum half-turn pulses, i.e., the channel identification numbers of any track are rigidly fixed in position relative to the half-turn pulse. The drum utilizes the main pulse sequences with the introduction of proper delays to form the control and gating pulses.

The channel-by-channel storage of data is realized by a logic device (the number of such devices is equal to the number of operational tracks), which permits two basic modes of operation:

a) The "add one," or "+1," mode.

In this case the operation of scanning for "first zero" is executed, followed by rewrite of the reamining positions of the number in the channel without alteration.

b) The "subtract one," or "-1," mode (Fig. 4 Tu1).

Here a "first one" scan is executed directly, followed by rewrite.

The actual channel logic network of each track comprises two triggers, six diode-transformer gates, and one transformer anticoincidence circuit. In terms of space economy it is best to combine this subsystem in a single package with the output-gated read amplifier, write amplifier, and certain other elements essential to the alignment and monitoring of the overall operation of the block. A diagram of the latter is shown in Fig. 4.

The phased input pulse actuates the logic trigger 1. Gates K_1 and K_2 are then opened. Through the open gate K_1 the actuating pulse energizes the phasing trigger 2, which opens gate K_6 to the write-ones position. If a pulse occurs at the output of the gated read amplifier (meaning a one has been written in the given digital position of the channel number), this pulse acts through the open gate K_2 to return trigger 2 to the position corresponding to open gate K_5 and closed gate K_6, i.e., the zeros-write or erase position. The erasing operation is continued as long as a read pulse is present at the output of gate K_2. The moment there is no pulse at the output of gate K_2 trigger 2 stays in the write-ones position, K_6 remaining open. This corresponds to the operation of scanning for "first zero." A one is written in its place by means of the nearest write gate pulse. Simultaneously this pulse acts through a feedback line (FB) to return logic trigger 1 to its initial state, at which time gates K_3 and K_4 are open. In this case the remaining positions of the number carried by channel N are rewritten without alteration.

In the "subtract one" mode the feedback circuit FB (see Fig. 4) is tied in with the output of gate K_5, corresponding to the write-zeros (erase) position. The operation of scanning for "first one" is executed, and zero is "written" in its place, followed by rewrite.

The anticoincidence circuit AC (Fig. 4) inhibits the switching of logic trigger 1 to the rewrite position when the "±1" operation performed earlier for the leading position (2^{13}) of channel N must be repeated for any position of channel N + 1, i.e., when a registered pulse is also transmitted to channel N + 1.

The pulses to be registered are phased in by the pulses of the channel sequence, which are also pulses for clearing the logic trigger 1. Clearing to "0" or erasing of the accumulated information is realized by interruption of the write-ones circuit (contact p_1 opens the write gate circuit). Relay R is used for this purpose. Provision is made for remote clearing and clearing during alignment operations by means of a duplicate switch Tu3.

The switch Tu2 is incorporated for testing of the channel logic block. In the "test" position writing is executed in all channels. The registration pulses in this case are the pulses of the channel sequence.

The read amplifier, complemented by the common serial voltage feedback loop, consists of two amplification stages and has a transformer input (n = 2). The general expression for the voltage gain K_{FB} of the amplifier has the form

$$K_{FB} \approx \frac{K_u}{1 + \dfrac{R_{e2} K_u}{R_{\kappa 2} n}} ,$$

where K_u is the voltage gain of the amplifier without feedback and n is the transformation ratio. In our system it is sufficient to have K_{FB} = 120-150. The block is installed in printed circuit form on a plate 90 × 480 mm.

LITERATURE CITED

1. I. V. Shtranikh, V. N. Bochkarev, A. N. Volkov, A. M. Klabukov, and V. V. Puzanov, Trudy FIAN, 42:3 (1968). [This volume, p. 1]
2. L. P. Kraizmer, Discrete-Information Storage Devices, Gosénergoizdat (1961).
3. I. V. Shtranikh, V. N. Bochkarev, A. N. Volkov, and A. M. Klabukov, Proc. Fifth Sci.-Tech. Conf. Nuclear Radio Electronics, Vol. 2, Part 2, Gosatomizdat, Moscow (1963), p. 135.

A SIMPLE BLOCK FOR BISECTION OF THE "MEMORY" OF THE AI-100 ANALYZER

A. N. Volkov and A. M. Klabukov

It is advantageous in a number of situations to be able to perform measurements on the AI-100 analyzer in one of two modes: a) with a straight 100 active analyzer channels and a single sensor to pick up the experimental pulses; b) with 2×50 channels and two sensors.

A schematic of the accessory by which this is accomplished is shown in Fig. 1. Pulses from the two divided inputs are mixed through diodes D_1 and D_2 on resistance R and are transmitted to the common input of the AI-100. Those pulses which arrive at the second input are transmitted simultaneously to the "memory" bisecting network. If the given pulse is a candidate for analysis (its amplitude being larger than the lower threshold of the AI-100) and the analysis of the preceding pulse has terminated, a "+1" increment pulse is generated in the accessory and is transmitted to the first and third cells of the tens decade of the AI-100 distributing device. In this way the number 50 is entered into the address device well ahead of the beginning of the amplitude-to-time conversion process, and the experimental signal can only be

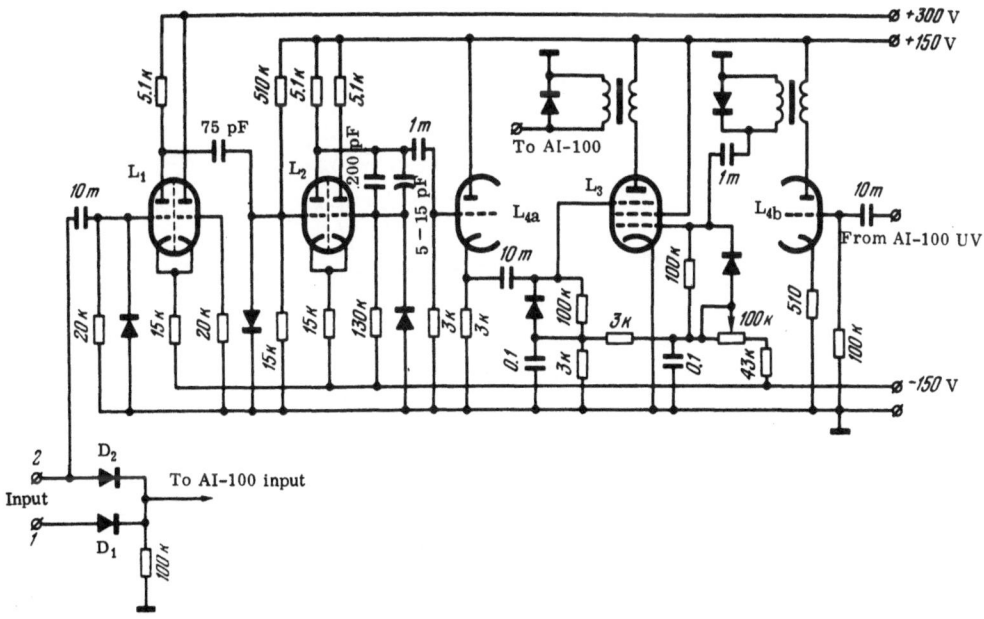

Fig. 1. Schematic of the AI-100 "memory" bisecting network. All diodes are type D2E, except D_1 and D_2.

registered in the range from the 51st through the 99th channel. The increment pulse is sent to the address device only in the event two pulses arrive simultaneously at the coincidence circuit L_3, viz., a pulse formed from the input signal through the second input by means of the univibrator L_2 of the accessory network, and a pulse from the univibrator UV of the analyzer input section (see the description of the AI-100). The threshold of the accessory univibrator L_2 is set somewhat below that of the AI-100 univibrator UV. This makes for uniform analyzer characteristics with respect to both inputs and eliminates the registration of pulses from the first input in the channels associated with the second input.

In order to preclude errors due to discrepancies in the registration of large pulses, the blocking univibrator (see the description of the AI-100) of the analyzer control device admits, in addition to the pulse from the 100th channel, the pulse from the 51st channel. The arrival of the increment pulse in the address device also causes the blocking univibrator to be actuated, but this induces a certain loss on the part of the experimental pulse, because the arrival of the increment pulse does not concur in time with the onset of the read-write process.

The network is supplied by the power supply section of the AI-100. The accessory contains four tubes, is mounted on a narrow Duralumin strip (20×8 cm), and is housed in the lower compartment of the analyzer. Not once in two years of service was there failure on the part of the network, and it exhibited uniform performance characteristics with respect to both inputs.

TUNNEL DIODE OVERLOAD-PROTECTION NETWORKS FOR SEMICONDUCTOR VOLTAGE STABILIZERS

A. N. Volkov and M. I. Salokhin

The principal shortcoming of relay-type overload networks is their slow firing time. There are several types of transistorized overload networks with a fast response time, but they are not capable in general of complete cutoff of the transistors or the realization of zero potential at the output in the event of overloading.

Our proposed overload-protection network (Fig. 1) using a tunnel diode D_1 has an almost-instantaneous response (to 10^{-8} sec) and permits complete blocking of the output transistor. The network is simple and contains a minimum of components. It comprises a tunnel diode potential trigger D_1, triode transistor gate T_1, standard resistance R_s, resistances R_2, R_3 for setting the tunnel diode in its initial state, and capacitances C_1, C_2 governing the response time of the overload network.

The network functions as follows. The load current, which creates a potential drop across the standard resistance $U_s = I_L R_s$, shifts the operating point toward the peak current of the tunnel branch. When the load current attains its maximum admissible value I_{max}, the trigger is actuated, opening gate T_1, whereupon the collector of the output triode of the stabilizer balancing amplifier (T_2, T_3) is shunted by triode T_1, its potential becomes positive, and the bases of triodes T_4, T_5, T_6 are closed. In high-voltage stabilizers it is advisable for more reliable triggering of the output transistor to supply the emitter loads of transistors T_2, T_3, and T_4 from a source of positive bias. The overload response time is determined in the first approximation by the time constant τ:

$$\tau = \left(\frac{R_3 R_4}{R_3 + R_4} + R_2 \right) C_2$$

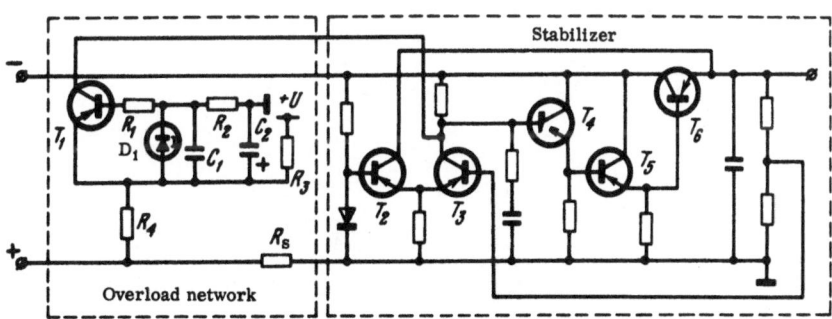

Fig. 1. Current overload-protection network.

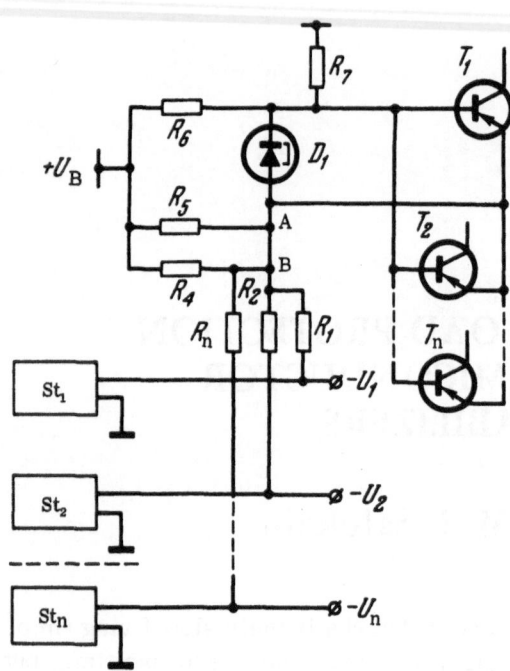

Fig. 2. Overload network for a multichannel voltage stabilizer. St_n) nth stabilizer; U_B) bias voltage.

and can be chosen within a wide range of values. For the given network (Fig. 1) the response time is 0.5 msec.

In cases when several power supplies are used (U_1, U_2, . . . , U_n) it is useful to have an overload network such that in the event of failure of one or more power supplies the output transistors of all stabilizers (T_1, T_2, . . . , T_n) will be cut off. A suitable network is shown in Fig. 2. Its operation is similar to that of the network in Fig. 1. Its main difference is that it is actuated by the failure of any stabilizer, rather than by simple current overload. In the initial state, if the line AB is broken, the potential of point B becomes equal to the potential of point A by a proper choice of resistances R_1, R_2, R_3, and R_4,

A SEMIAUTOMATED DEVICE FOR MEASURING THE GEOMETRIC PARAMETERS OF CHARGED-PARTICLE TRACKS IN PHOTOEMULSIONS

A. E. Voronkov, G. E. Belovitskii, L. N. Kolesnikova, V. S. Marenkov, M. F. Solov'eva, L. V. Sukhov, and P. N. Komolov

Introduction

Lately, in order to increase the speed and accuracy of measurements of the parameters of charged particle tracks in nuclear photographic emulsions, considerable importance has been attached to semiautomated scanning and measuring devices. A great many such devices have been designed and built [1-10]. Usually they comprise a standard or specially fabricated instrument microscope and specialized ancillary equipment. Other special equipment is used to convert the instantaneous coordinates of the microscope platform into digital numbers. Sometimes the platforms are equipped with drive motors. Electronic devices utilizing perforation or magnetic recording techniques transform the acquired information into a form suitable for input to digital computers. In some case the required computations are performed in the instrument itself, and the readout of data is accomplished by means of digital printers [4, 6, 8]. The participation of the operator in semiautomated devices is limited to visual inspection of the photoemulsion and tracking of the particles by manual or motor-driven control.

In 1963 a semiautomated device was designed and built in the Atomic Nucleus Laboratory of the Institute of Physics of the Academy of Sciences (FIAN) of the USSR for measuring the geometric parameters (ranges and angles) of particle tracks in a photoemulsion (Fig. 1). This device was used to measure the energy spectra of scattered neutrons with initial energies of 14 MeV. The system comprises a microscope with a device for converting its positional coordinates into digital form, and an electronic and mechanical subsystem for motor-controlled tracking, the execution of various arithmetic and other operations, and the output of measurement data onto punched cards.

1. The Microscope

The mechanooptical part of the device is constructed mainly on a conventional microscope, type MBI-3, with a binocular attachment. The coordinate-tracking platform of the microscope is a cross-shaped platform that can be displaced in two mutually perpendicular directions (X and Y). The platform is displaced along glass guides on ball bearings. The platform is equipped with a vernier scale and has an adjustable 360° turn mechanism above the two guides. Stepwise displacement of the platform is executed by means of step motors with step intervals of 1 and 0.7 μ in the X and Y directions, respectively. The magnitude and direction of the step

63

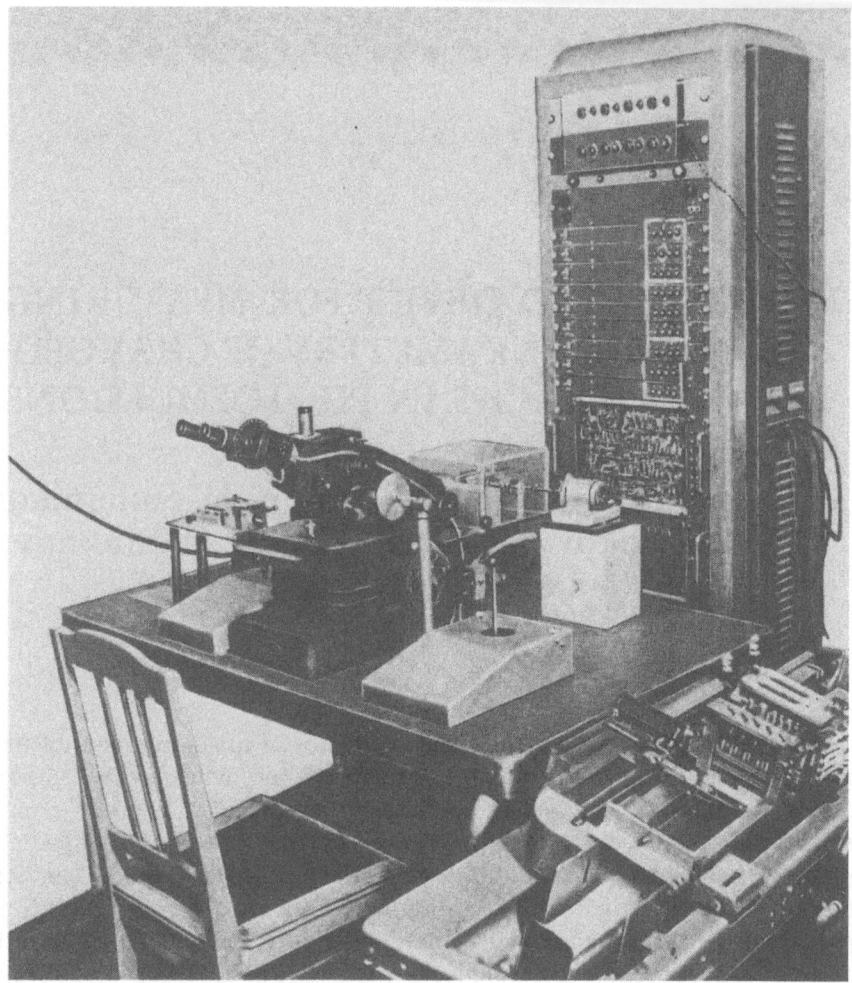

Fig. 1. Overall view of the equipment.

motor outputs are controlled by a special goniometric device attached to a control lever. This device facilitates and speeds up the tracking operation. The platform is set in return motion by weights (400 g) located under the main slab of the platform.

Displacement in the Z direction is executed manually. The digitalization of the Z coordinate is realized by the installation on the microscrew knob of a differential photoelectric pulse pickup developed at FIAN. The pickup is attached to the knob of the microscrew by a gear drive with a 1:3 ratio, giving a scale division on the Z axis of $0.33\ \mu$. In order to achieve the required measurement accuracy in the Z direction, it is necessary to perform measurements of the depth of the ends of the track, focusing at all times toward the surface of the emulsion. This makes it possible to neglect the intrinsic backlash of the focusing system.

For scanning of the emulsion a 60× objective with an aperture of 1.0 is used, along with K10× oculars, one of which has inserted in it a scale (Fig. 2.) with a superimposed angle defining the tracks to be scanned and an etch mark with a break in the middle $2\ \mu$ in the length (when translated to the objective). The points of the track whose coordinates are to be measured fall inside the break. A portion of a circle $7-8\ \mu$ in diameter is centered between the etch lines on the scales. The diameter of the circle defines the maximum length of the sharply visible portion

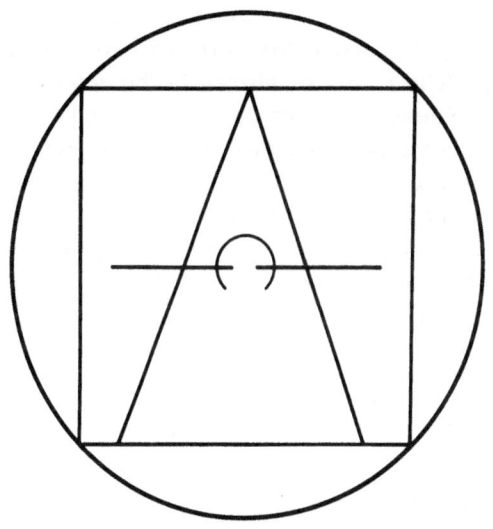

Fig. 2. Ocular scale.

of the track for proper selection of the tracks with respect to angle of immersion in the emulsion. Moreover, two vertical and two horizontal lines are superimposed in the field of view in order to delimit the scanned area of the field of view. The microscope is illuminated with a type OI-19 lamp attached to the underside of the platform.

The mechano-optical part of the system was assembled from available parts and can be designed for other components. For example, a type MBI-9 microscope can be used with this kind of device after a few minor modifications.

2. Mechanoelectronic Part

of the System

This part of the system facilitates the motor-driven control of the X and Y coordinates, accumulates and converts into digital form the information on the three coordinates, and transfers this information to punched cards by means of a modified P-80-6 card punch. In addition, the device permits automatic return to the field of view in which tracking of the given track was initiated. A block diagram of the device is shown in Fig. 3.

The goniometric device (Fig. 4), which is attached to a control lever, moves the rotors of two wire-wound resistors at an angle of 90° to one another according to the inclination of the control lever. The end points of these resistors are connected to a 400–cps two-phase alternating potential source. By suitable design measures it can be arranged so that, with the lever inclined in any direction up to about an angle of 45° from the vertical, voltages will be taken from the rotors of the potentiometers, one almost proportional to the sine, the other almost proportion to the cosine of the angle between the selected direction in the horizontal plane (normally it coincides with the X axis) and the projection of the lever onto that plane (see Fig. 4). The direction to the projection of the lever coincides with the direction of the track during tracking. The maximum voltage amplitudes will increase with the angle between the vertical and the direction of the lever itself. Each direction is admitted to the input of an amplifier, which supplies the control phase of a type DG-0.5A two-phase motor with a built-in tachogenerator, providing strong tachometric feedback in the amplifier–motor system. As a result, the speeds of rotation of the motor follow the input voltages of the amplifier with high precision down to very slow rates of speed (fractions of an rps). Lightweight disks are attached to the axles of the DG-0.5A motors to form, in combination with photodiodes, illuminators, and appropriate electronic circuits, phase-sensitive frequency-dependent pulse transmitters.

The device just described, including the amplifier, motor, and photocell with associated electronic circuitry, actually represents a pulse generator with an extremely linear frequency range. Two such generators operating in conjunction with the electromechanical control device described above (control lever) form a complex two-channel network that generates voltage pulses at two different frequencies [11]. The voltage pulses at the two frequencies drive type DSh-0.1 coordinate step motors, which operate through backlash-free reducers to transport the microscope platform in the direction of the projection of the lever on the XY plane at a speed proportional to the slope of the lever. These frequencies are admitted simultaneously into appropriate X and Y binary reverse counters, each with a capacity of 2048 bits.

Motion in the Z direction is executed by rotation of a microscrew, to which is attached a photocell D_z similar in design to photocells D_x and D_y. The pulse from this cell are transmitted to a binary reverse counter, also with a 2048-bit capacity. The numbers stored in the counters can be transferred to punched cards by pressing a button. These numbers can correspond either to end points or intermediate points of the track, and for this purpose the device is furnished with several transfer buttons. The system comprising the step motors and counters essentially fixes the distances traversed by the coordinate platform in the X and Y directions, as expressed in terms of the number of steps of the motor. This facilitates automatic return of the platform to the initial position. The precision with which the platform is returned falls within the limits of the scale division for one step, i.e., 1 μ on the X axis and 0.7 μ on the Y axis. The "automatic return" operation is carried out in the device by pressing a button B located in the casing of the control lever (see Fig. 4).

For convenience in aligning the selected positions of the tracks with the spacing in the etch mark in the middle of the field of view, the device is equipped with three pedals, one for simultaneous X and Y motion and two for separate displacements in the X and Y directions.

The device utilizes a P-80-6 card punch, which is designed for manual or keyboard operation. For this reason we redesigned the card punch for automatic operation in connection with the device. The modified design included certain additions to the electrical circuitry of the card punch, which were connected simultaneously into the circuitry of the device as a whole, along with a few minor mechanical modifications.

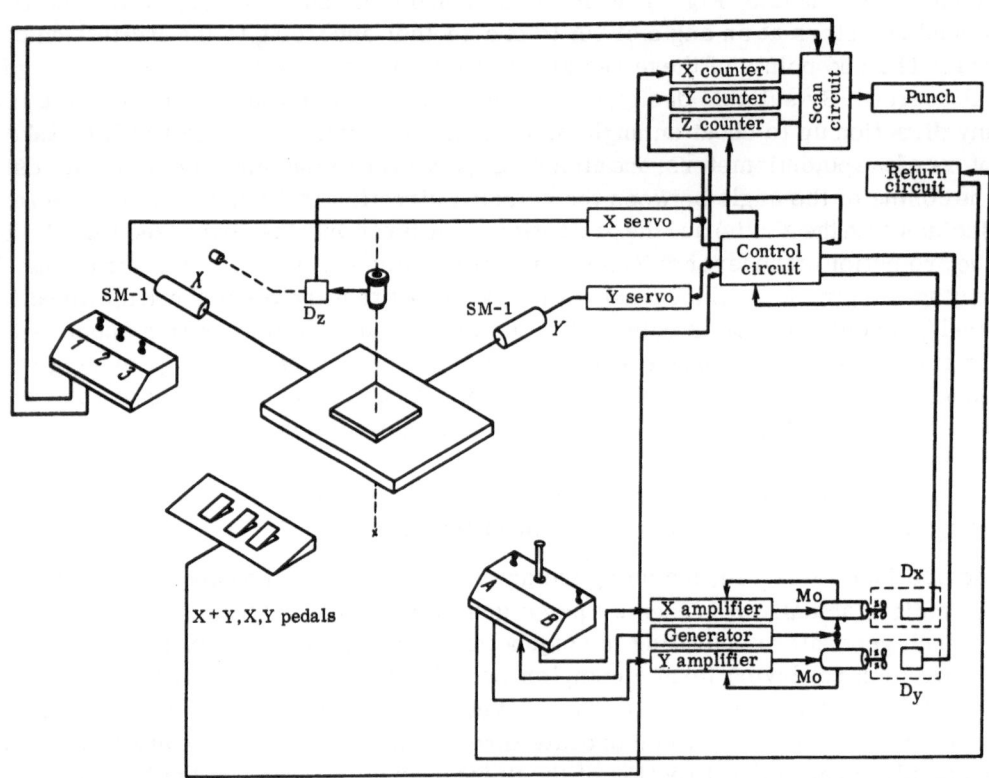

Fig. 3. Block diagram of the device.

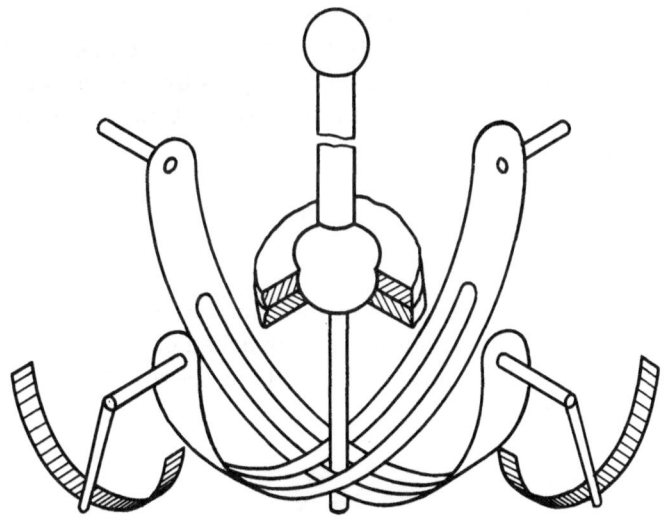

Fig. 4. Diagram of the goniometric device.

3. Scanning of Photographic Plates by Means of the

Semiautomated Device

The instrument described above has been used to measure the energy spectra of scattered neutrons produced in the interaction of 14-MeV neutrons with various nuclei. The neutron energy was determined by measuring the range of the recoil proton and the angle between the track of the recoil proton and the direction of the primary neutron. The recoil protons were registered by means of NIKF (Motion Picture and Photography Scientific Research Institute) type "K" thick photographic plates with a film thickness of 400 μ. The ranges of the recoil protons attained 1000 μ, i.e., a distance covering several (five or six) fields of view. Consequently, the following procedure was developed for scanning and measuring the recoil proton tracks.

By means of a button A situated on the casing of the control lever, the observer sets the next field of view. Entering into this field of view is a proton track beginning in the emulsion and forming an angle of at most 15° with the direction of the primary neutron beam (in the horizontal and vertical planes). On detecting such a track, the observer depresses the foot pedal and with the control lever shifts the track at a speed of 50 to 100 μ/sec until it comes to its end. If the recoil proton emerges from the emulsion, the observer presses the return button, also located on the casing of the control lever, thus bringing the platform back to the initial position (correct to about 1 μ). If the recoil proton stays in the emulsion, the track is measured. The end of the track is set in the gap in the middle of the field of view and by means of button 1 the X, Y, and Z counters are taken into the zero position. Then the track is moved, and at a distance of ~80 μ from its beginning it is again aligned in the gap. By pressing button 2, the operator transfers to a punched card the lengths of the projections of the measured portion of the track on the X, Y, and Z axes. Then the beginning of the track is set in the gap, and by pressing the button 3 the operator transfers the analogous data for this portion of the track onto a punched card. The track terminates with this measurement. The platform is set in the initial position by means of the return button B. The observer continues to scan and measure new tracks in the given field of view, then proceeds to scan the next field of view, and so on. Normally in the measurements the track is divided into two parts, the data for each being transferred to one line of the punched card. If, however, the recoil proton has undergone scattering through a large angle, the total length of the track is measured in three, rather than two parts.

In any case the cosine of the solid angle between the track of the recoil proton and the track of the neutron is computed from the initial portion of the track (of length $\sim80\ \mu$). The error induced in the determination of the neutron energies by fluctuation of the angle between the neutron and recoil proton tracks as a result of multiple Coulomb scattering of the recoil proton is minimal.

4. Digital Computer Processing of the Punched Cards

The calculation of the energy spectrum of the neutrons from the measured recoil proton tracks, the data for which have been transferred to punched cards, is performed by a special program on an electronic digital computer. The neutron energy E_n is related to the recoil proton energy E_p by the expression

$$E_n = \frac{E_p}{\cos^2 \psi}. \tag{1}$$

Here $E_p = E_{p_0} + ar^b$, E_{p_0} is a constant depending on the type of emulsion used, r is the track length, a and b are coefficients found for a standard emulsion,

$$\cos \psi = (\alpha x + \beta y + \gamma Kz)/\sqrt{x^2 + y^2 + (Kz)^2};$$

ψ is the solid angle between the direction of the neutron track n and direction of the recoil proton track $p(\psi_0)$, x, y, z are the lengths of the projections (in microns) of the initial portion of the track, K is the reduction factor, and α, β, γ are the direction cosines of vector n.

The shape of the spectrum for a monochromatic flux of neutrons having energy E_0 is strongly dependent on the choice of parameters

$$\beta; \gamma; \varepsilon = E_0 - \frac{E_i + E_{i+1}}{2}; \quad K; E_{p_0}, \tag{2}$$

where E_i and E_{i+1} are the extreme points of the energy interval including E_0 on the histogram.

For the calculations the optimum values of (2) are selected automatically. Suppose that the neutron flux n is directed along the X axis (as in Fig. 5), so that $\beta_0 = 0$. Normally in experiments $\beta_0 \neq 0$. By making the invalid assumption that $\beta_0 = 0$, we impart to p_1, which has y > 0, an angle $\psi_1 > \psi_{01}$, i.e., $E_1 > E_0$. For track p_2, accordingly, $\psi_2 < \psi_{02}$, i.e., $E_2 < E_0$. Consequently, an incorrect choice of the parameter β distorts the spectrum.

In spectra such that the elastic scattering peak differs from the first peak corresponding to inelastic scattering by a fairly large energy interval, the optimum values of β and γ are computed by means of approximate equations that we derived on the basis of a statistical treatment of the problem. The equation for β takes the form

$$\frac{\sum_s y_{i-1,i}^{(s)}(\beta,0)}{N_{i-1,i}(\beta,0)} - \frac{\sum_s y_{i+1,i+2}^{s}(\beta,0)}{N_{i+1,i+2}(\beta,0)} = 0, \tag{3}$$

where $N_{i-1,i}$ is the number of neutrons in the energy interval $E_{i-1,i}$ and $\sum_s y_{i-1,i}^{(s)}$ is the algebraic sum of the y coordinates of the tracks corresponding to neutron energies in the given interval.

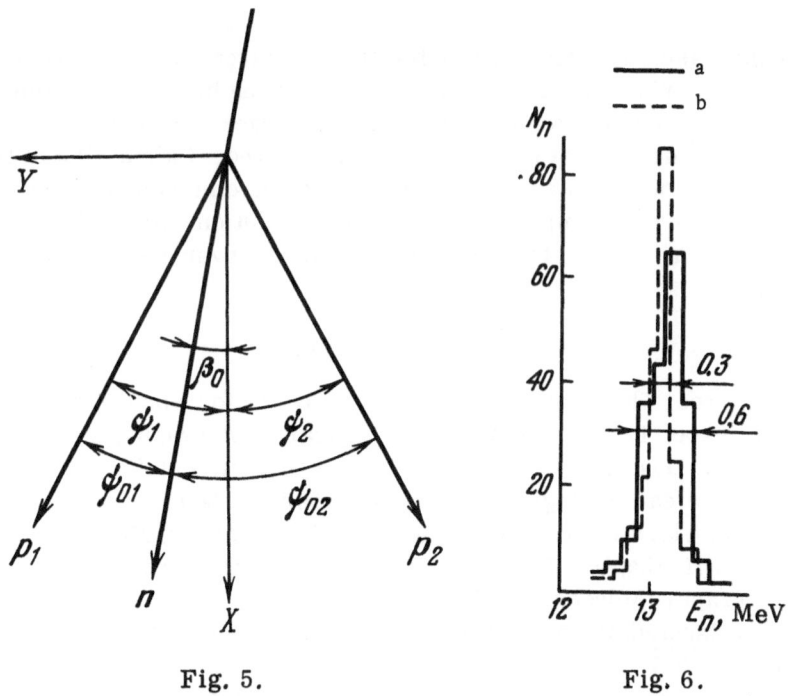

Fig. 5.

Fig. 6.

Fig. 5. Diagram pertinent to the derivation of Eqs. (3) and (4).

Fig. 6. Energy spectrum of a primary beam of 13.2-MeV neutrons.
a) Spectrum computed for $\beta = \gamma = 0$; b) the same, for optimum values
of β and γ.

The equation for γ is analogous:

$$\frac{\sum_s [Kz_{i-1,i}^s (\beta_0, \gamma)]}{N_{i-1,i}(\beta_0, \gamma)} - \frac{\sum_s [Kz_{i+1,i+2}^s (\beta_0 \gamma)]}{N_{i+1,i+2}(\beta_0, \gamma)} = 0 . \tag{4}$$

The equations have been solved by the method of linear interpolation.

In the more general case of poorly resolved elastic and inelastic scattering peaks the following is used as a criterion for selection of the parameters (2):

$$\max\{\max[N_n(\beta, \gamma, \varepsilon, E_{p_0}, K)]\}. \tag{5}$$

The quantities E_{p_0} and K are fixed for a given emulsion; the computer constructs the spectra by varying β, γ, and ε within the prescribed limits.

The energy spectrum of a primary beam of 13.2-MeV neutrons is shown in Fig. 6.

For proton tracks chosen within an angle of $\pm15°$ with respect to the direction of the primary neutrons the width of the curve at the half-peak points is equal to 0.3 MeV. Therefore, the relative error in measuring a neutron energy of 13 MeV is ~2%. The energy resolution obtained in these measurements is greater than the resolution attainable in measurements of neutrons of the same energy by the time-of-flight method, being equal to 0.45 MeV.

The following conclusions are made on the basis of a two-year experimental program using the semi-automated device. The device greatly increases the efficiency of the observer. This is due, first of all, to the fact that he is relieved of having to write down the measured values (range, height, angles); second, the displacement of the track by means of a single control lever greatly increases the rate of scanning and measurement of the track ranges; third, the observer is free to perform measurements without lifting his eye from the ocular, and this means he can scan the emulsion when a large number of tracks are present in the field of view without the danger of losing sight of the track being measured. Calculation of the neutron spectrum with all necessary corrections being made on an electronic computer also affords a significant time-saving feature and improves the energy resolution. As a whole, the device provides a five- or sixfold increase in the efficiency of the observer and enhances the accuracy of the measurements.

LITERATURE CITED

1. W. H. Barkas, Proc. Third Internat. Conf. Nuclear Photography, Moscow (1962), p. 317.
2. G. R. Holbeck, Proc. Third Internat. Conf. Nuclear Photography, Moscow (1962), p. 327.
3. R. E. White, A. E. Dyson, S. C. Freden, H. N. Kornblum, and J. R. McCall, Proc. Third Internat. Conf. Nuclear Photography, Moscow (1962), p. 340.
4. M. J. Duff, Proc. Third Internat. Conf. Nuclear Photography, Moscow (1962), p. 348.
5. F. W. O'Dell, M. H. Shapiro, B. Stiller, and R. C. Waddel, Proc. Third Internat. Conf. Nuclear Photography, Moscow (1962), p. 357.
6. C. Gal't'eri, A. Manfredini, and V. Rossi, Proc. Third Internat. Conf. Nuclear Photography, Moscow (1962), p. 367.
7. É. G. Melikyan, Proc. Third Internat. Conf. Nuclear Photography, Moscow (1962), p. 376.
8. W. H. Barkas, H. H. Heckman, G. Hodges, and J. G. Salvador, Korpuskularphotographie [Particle Photography], Vol. IV, Munich (1963).
9. P. W. Benjamin and G. S. Nicholls, UK Atomic Energy Authority AWPE Report NR5/63 (1963).
10. K. Breuer and E. Rossle, Fifth Internat. Conf. Nuclear Photography, CERN, Geneva, September 15-18, 1964.
11. I. V. Shtranikh and A. E. Voronkov, Advanced Scientific, Engineering, and Industrial Practices, No. 9, TsITÉIN, Moscow (1961).

TIME ENCODER FOR THE MEASUREMENT
AND REGISTRATION CENTER

I. V. Shtranikh, A. M. Klabukov,
and A. E. Samsonov

One of the measuring circuits serving the laboratory measurement and registration center is designed for pulsed-mode neutron spectrometer measurements of the slowing-down time in lead. The relative time resolution is approximately 20% [1].

Under these conditions the measurement interval of the time encoder must be subdivided into groups having different channel widths [2], in order to maintain a constant channel resolution for small times (at the beginning of the measurement interval of the encoder).

The time encoder is designed to convert time intervals into binary code and to prestore that code (first buffer memory). The position of the time interval-distance is chosen by preliminary setting prior to the beginning of measurement.

There are 128 channels, which for simplicity are divided into four constant groups with different channel widths: 32 channels at 4 μsec, 32 channels at 8 μsec, 32 channels at 16 μsec, and 32 channels at 32 μsec.

The encoder has four blocks (Fig. 1), including: I) a phasing block (with 4-μsec phasing pulses); II) a block for automatic switching of the channel groups and the logical operation of generating spacing (delay); III) a time-counter block for conversion of the time intervals into binary code and the generation of spacing; IV) a buffer memory block. Provision is made for prestorage of the code and phasing of code readout at the timing frequency of the buffer memory (derandomizing-type memory).

The "dead" time is at least 4 μsec and depends mainly on the response time of the derandomizing memory.

The initial state is as follows. All the gates are closed, and the triggers are in the "0" position. On receiving a START signal, Tr1-Tr7 of block III are "cleared" to "0," K_6 and K_7 are opened, and a 4-μsec sequence is transmitted to the time counter to set the spacing (delay). After readout of the spacing the signal closes K_6 and K_7 through S_1, K_7, and Sh_2 of block II, sets the time counter in the "0" state, and opens K_2 of block II and K_1 of block I. Through K_2 of block II a 4-μsec time marker begins to arrive at the time counter. With the arrival of the 32nd pulse a signal from Tr5 of block III switches the channel width to 8 μsec through Sh_1 of block II by means of a conventional trigger-type shift register, etc. With the arrival of a signal from the detector Tr1 of block I closes K_1 and opens K_2 of block I. The first phasing pulse passes through Sh_1 of block I to read the code from block III, transferring it to buffer memory. Reading is executed without the time counter stopping. Trigger Tr1 of block IV is used for phasing in with the timing frequency of the buffer memory.

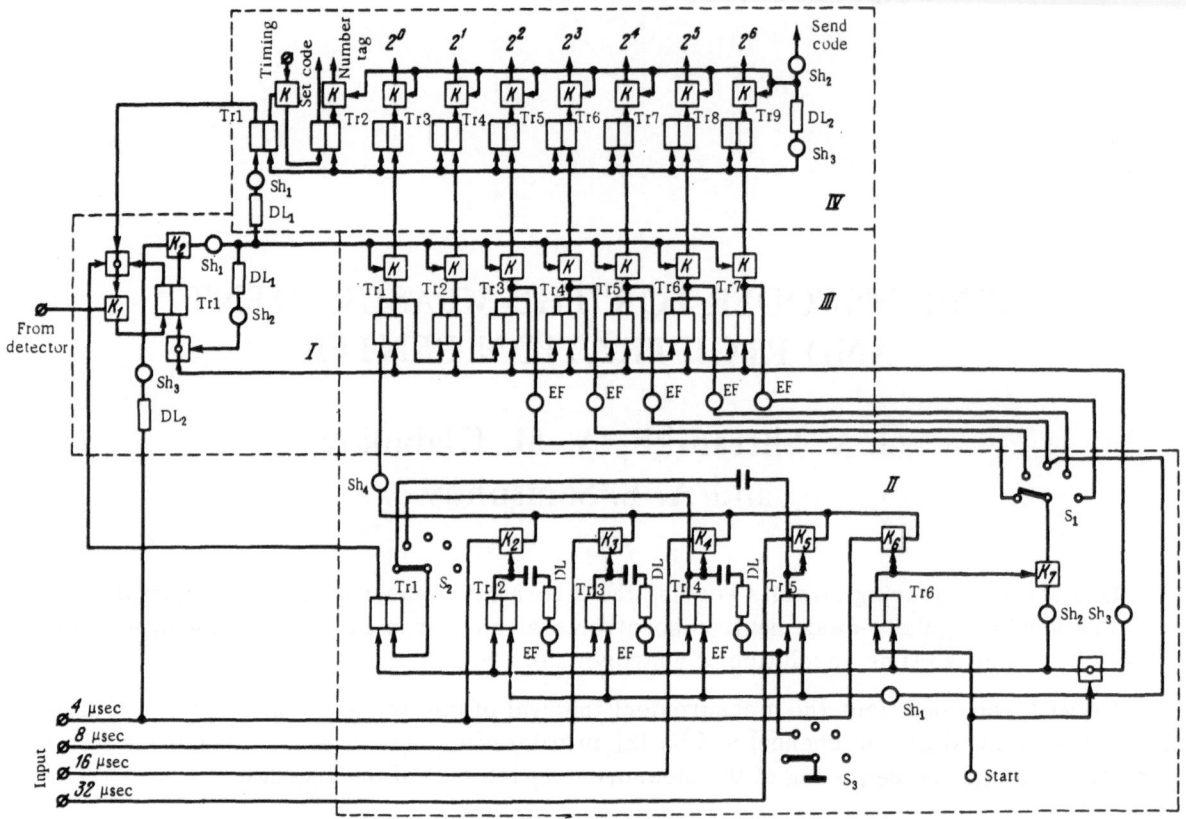

Fig. 1. Functional diagram of the 128-channel time encoder with subdivision into four equal groups of channels with different channel widths (4, 8, 16, and 32 μsec).

A block diagram of another time encoder with the same partitioning of the 128 channels into four groups, but with channel widths corresponding to 0.25, 0.5, 1, and 2 μsec, is shown in Fig. 2. The basic difference between this and the preceding encoder lies in its three parallel coding branches and its ability to accept three pulses during each operating cycle of the physical apparatus, i.e., it functions as three encoders with a single input, but controlled by a common logic network.

The parallel encoder has four blocks, including: I) a phasing block; II) a channel-width switching block; III) a block of address counters and three-word buffer memory; IV) a logic block for output of the codes into a derandomizing-type buffer memory.

On reception of the START #1 signal, the time counters are cleared to "0" through K_{22}, K_{23}, and K_{24} on the condition that Tr5, Tr6, and Tr7 are in the "0" state. Tr9 closes K_{25}, thus stopping the output of codes. After a period of time equal to the cycle of the derandomizing-type memory the START #2 signal arrives. At this instant measurement of the time interval to arrival of a pulse from the detector is initiated. After readout of 32 marker pulses of a 4-Mc sequence, a signal from trigger 2^4 for code #3 switches the series to 2 Mc through Sh_2, and so on.

The first detector pulse takes Tr7 into the "1" state. Gate K_{10} is opened, and " code #1" remains in the triggers. Gate K_9 is opened simultaneously. The encoder can admit three pulses separated by time intervals of 0.25 μsec. If three pulses arrive prior to the next cycle, reading of the codes is initiated through the OR_{15} circuit. Trigger Tr9 and gates K_{25}, K_{26}, K_{21} form a phasing network on the timing frequency of the derandomizing memory. Shaper Sh_{11} is used

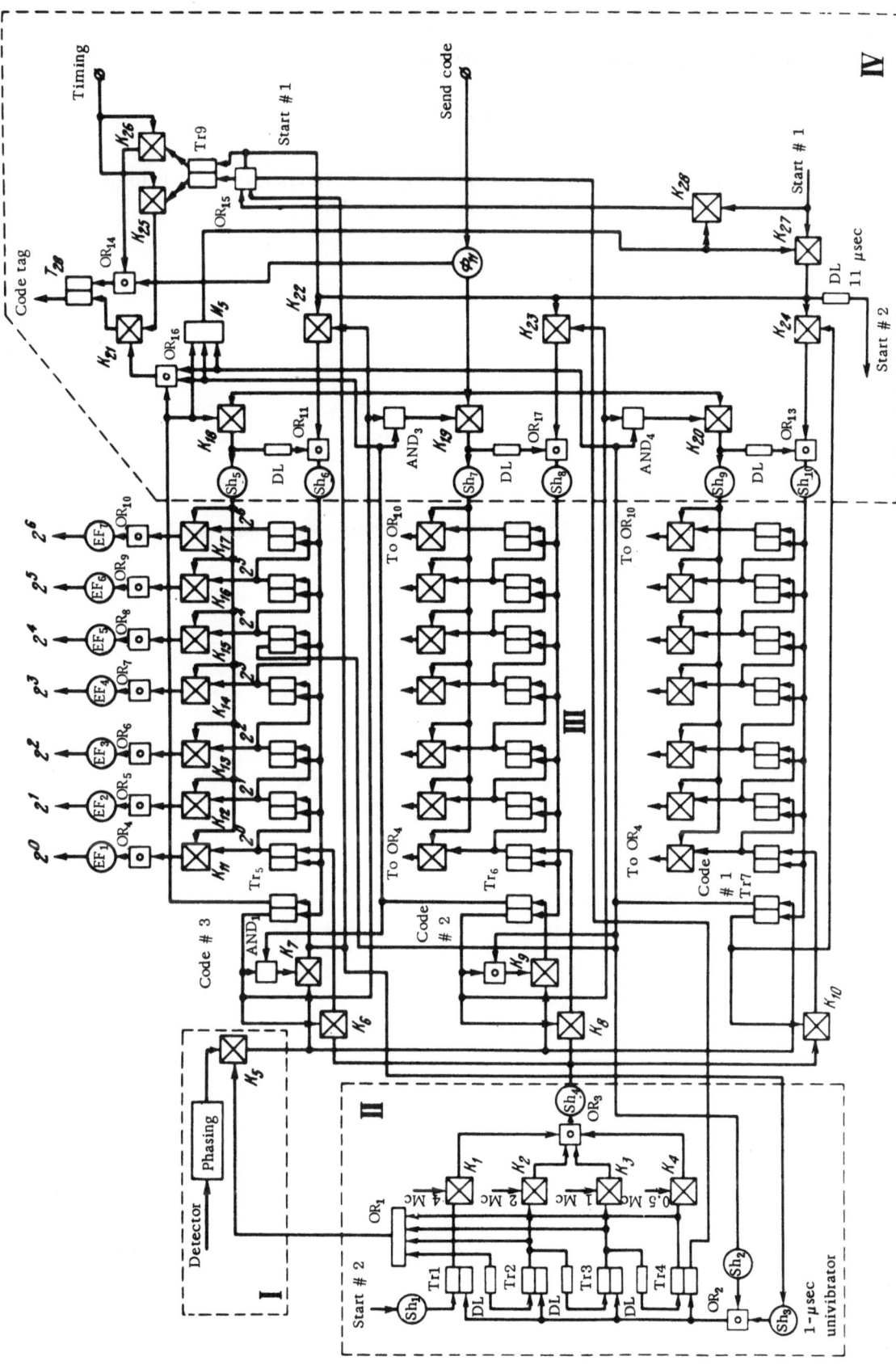

Fig. 2.   Functional block diagram of a time encoder with parallel branches.

for output of the code and clearing of the address counter triggers to "0." Output of the codes begins with "code #3," but if the latter is absent output begins with "code #2," etc. This is accomplished by means of gates K_{18}, K_{19}, K_{20} and the AND_3, AND_4 circuits.

LITERATURE CITED

1.   Research on Neutron Physics, Trudy FIAN, 24 (1964).
2.   I. S. Krasheninnikov, S. S. Kurochkin, E. I. Pekhin, V. V. Eldashev, R. S. Efimchik, A. S. Tuchina, Trudy SNIIP, No. 1, Atomizdat (1964), p. 79.

USE OF A CARD PUNCH FOR THE
ROW-AT-A-TIME INPUT OF DATA

A. M. Klabukov, A. A. Rudenko,
and I. V. Shtranikh

It is always desirable in laboratory practice, from the viewpoint of reducing the bulk of registering instrumentation, to amplify the functional capabilities of the various instruments in the latter. One specific approach is to convert standard type-P-80 card punch machines to operation in a row-at-a-time command input mode using keyboards or other devices.

Two versions of such a remodeled machine are described below. One version executes phased row-at-a-time input to the output card punch with continuous operating of the punch motor. The punching speed has a maximum value of 10 lines per second. In the other version, which operates more slowly, the motor of the input punch is switched off after each row is punched.

Implementation of the first modification must meet the following specifications:

a) single punching of one row at a time, irrespective of the command arrival time or its duration;

b) constant duration of the current signals to the punch electromagnets, irrespective of the command arrival time or its duration;

c) row-to-row transport of the card in the matrix after each row is punched;

d) delayed transport of the next card until completion of punching of the 12th row of the card in process, followed by delivery of the new card under the first-row punch.

The conversion network is shown in Fig. 1. The card punch is equipped with an extra cam with standard contact. Relay A is designed for remote conversion of the system circuits from conventional continuous output to delayed batching operation.

Conditions "a" and "b" are realized in the operation of the card punch by timing pulses. This is done by inclusion of the conversion relay C, which is controlled from contacts b_1 and b_2 of relay B. Relay B duplicates the operation of the card punch pulse contact D_4 (for which D_4 is opened from D_6). In this case relay C converts the pulses from contact D_4, permitting the circuits of the system to be set for operation. An additional "weighted" power source comprising the capacitor and diode elements C_1 -C_3 and D_1 -D_2 is provided to operate relay C.

Contacts c_1 and c_2 of relay C set the discharge circuits of capacitors C_4 and C_5 in sequence through the winding of relay C. Since the discharge current from the capacitors has the opposite potential, the core of relay C is either magnetized through contact c_1 and the relay inhibited through contact c_2, or the core is magnetically reversed and thus released.

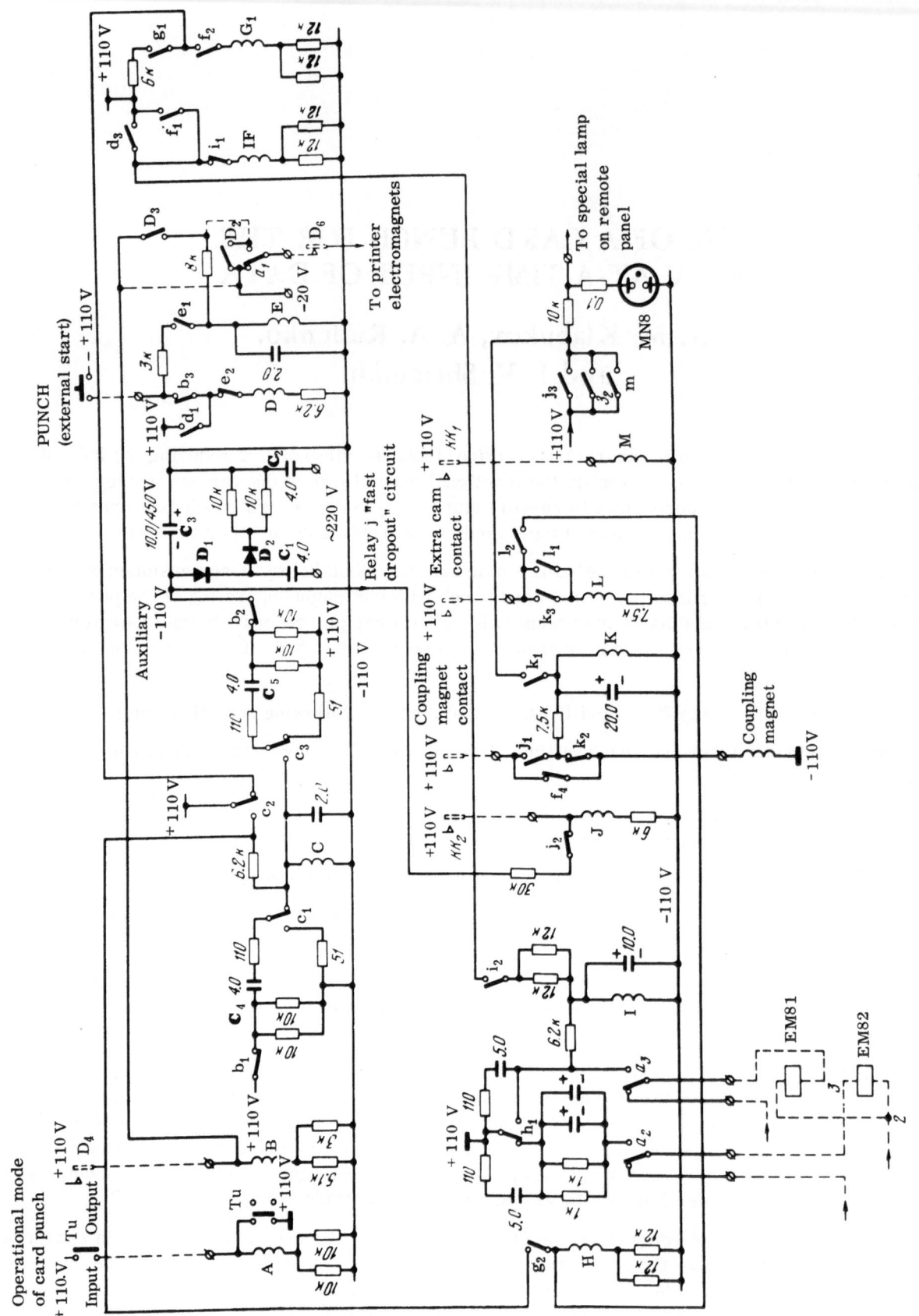

Fig. 1. Network for conversion of the output card punch to operation in a delayed data–input mode.

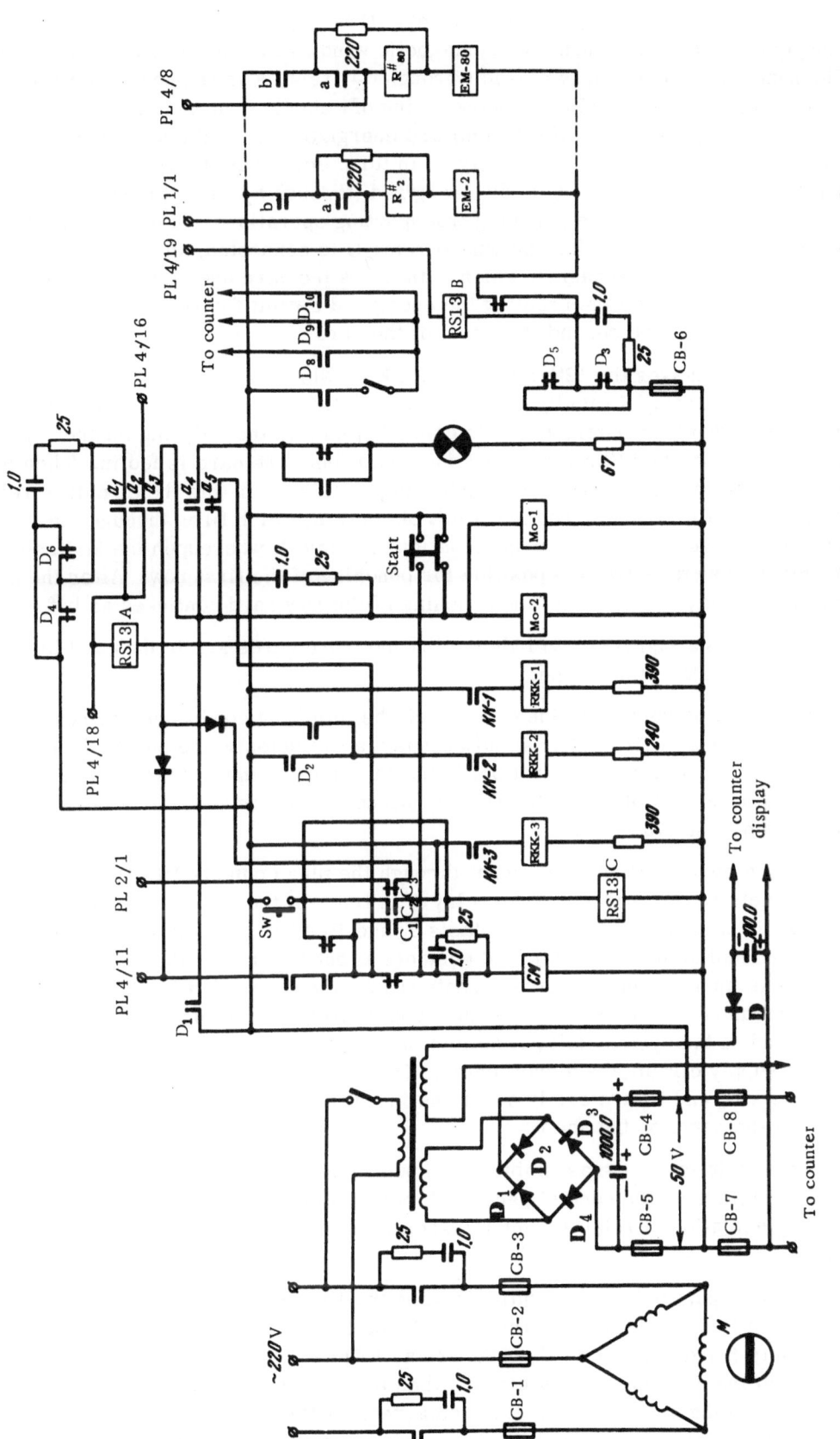

Fig. 2.   Control network of the input card punch for row-at-a-time operation in the command mode.

On receiving the external PUNCH command, relay D is tripped (with the cam contact D_4 open) and the supply circuit of the punch electromagnetic windings is energized (through d_2 to contact D_6). The next time contact D_4 is closed (through d_3), relay E is tripped with a time delay, whereupon it simultaneously inhibits a second firing of relay D (punch) in the event of a protracted PUNCH command (inhibited through e_1) and energizes the card transport circuit for the next step (tripping relay F through e_3). Relay F, in turn, energizes the supply circuit of relay G (through f_2), which trips only if C is de-energized (through c_2), and this subsequently provides proper separation of the card punching transporting operations. Moreover, relay G energizes the circuit of relay Hd which fires only if relay C is connected, and through contact h_1 actuates a Maltese-cross electromagnet, which transports the card one step. Then relay I is tripped with time delay, thereby interrupting the energized circuit for relay H (through contact i_1) and inhibiting a premature second transport of the card.

As soon as the card passes the 12th step, contact KK_2, which has a special small bias for this circuit, is opened, and relay J rapidly drops out, since its current polarity is reversed through contact j_2 and a 30-kΩ resistor, while its contact j_1 cuts in the coupling magnet, which drops out when relay K is tripped (through k_2). As a result, only one card is fed in. Then the card must be transported to the first row of the punching matrix. For this the circuit of relay K operates through contact k_3 to energize the circuit of relay L, which fires through the extra cam contact and again provides for long-term actuation of relay H, whereupon the Maltese-cross electromagnet moves the card into position for punching of the first row. After the extra cam contact is broken, relay L drops out, and punching of the new card continues as before.

Relay M functions as a signalizer to indicate that no more cards are in the card loader. All relays are of the normally-closed type.

In the second operating version of the card punch the following conditions are realized: a) single punching of a row at a time, after simultaneous transmission to the punch electromagnets of pulses no more than 10 msec in duration; b) power supply to the card punch entirely from the 220-V ac line; c) conversion of the motor from continuous to command operation. A circuit diagram of this version is shown in Fig. 2.

Relay A is closed by a short PUNCH signal through the plug contacts PL4/18 with inhibiting through cam contacts D_4, D_6, and a_1. When relay A is tripped, the following operations take place: A scan signal is transmitted through contacts a_2 and PL4/16 to the information sources, contacts a_3 cut in the coupling magnet (CM), and contacts a_2 cut in a motor (Mo-1 or Mo-2). When this happens, the marker punch electromagnets (PL2/1) are also tied in. The cam contacts D_4 and D_6 are mounted on the shaft so that the supply circuit is cut off after punching of the card row at the end of the cycle. Relay A cuts in, and all the circuits connected to it and the coupling magnet supply circuit are disconnected. The beginning and duration of cutoff of the circuits of contacts D_4 and D_6 are easily set in the initial adjustment of the card punch by suitable alignment of the cams on the shaft.

A separate panel allows for the assembling and punching of the control information. Incorrectly chosen characters on the control panel and, hence, in the magnetic block of the card punch are cleared by actuation of relay B, which disconnects the common supply circuit of the electromagnets.

Feeding of the first punched card is initiated when the FEED CARD button is pressed on the control panel. This ties in and inhibits relay C. Contacts c_2 cause a potential to be transmitted through contacts 5 and 6 of the card contact relay RKK-2, which senses the presence of cards in the matrix, to the coupling magnet circuit, then through contacts a_5 the motor is turned on. The supply circuit of the marker punch electromagnets (contacts c_3) is cut out simultaneously. As soon as the first punched card is positioned underneath the punch for punching of the first row and the card contact of the KK-2 matrix is closed, contacts 5 and 6 of relay RKK-2 are opened, inhibiting of relay C is discontinued, and the motor with the coil of the coupling magnet is shut off. The expulsion of a defective punched card is brought about by means of the START button on the card punch control panel.

The punch electromagnets are triggered by millisecond pulses through the start relays $R_2 \ldots R_{80}$, which are connected in series with each punch magnet. The current pulses from the information source fire these auxiliary relays and, on becoming inhibited through contacts a and b, cut in the proper electromagnets. If the time for actuation of all the type RSM-1 relays used for this circuit is equal to 3-5 msec, the card punch is reliably operated by 30-mA pulses of less than 10 msec in duration.

The relays and electromagnets are deenergized by breaking of the cam contacts D_3-D_5 as in the preceding network.

The maximum punching speed is not limited by the electrical network of the card punch, and with the motor in continuous operation the rated speed of the given card punch is maintained (two rows per second). Command operation of the motors is possible by virtue of the fact that the write commands are transmitted in the particular apparatus used at a speed much slower than one row per second.

MONITORING AND TIME-MEASURING CIRCUITS

A. M. Klabukov, A. E. Samsonov, and I. V. Shtranikh

The registration of auxiliary parameters, such as the intensity of radiation sources, astronomical time, the "live" and "dead" time of encoders, etc., is a prerequisite to any quantitative spectrometric measurements in nuclear physics. Without these measurements, which we call monitoring, it would be impossible to correlate and process the various results of physical experiments.

For the MRC system the problem of registering the secondary parameters is solved by fairly simple means. For every common group of tracks on the magnetic drum designed for one measurement another track is set apart for registration of the indicated parameters. In order to increase the channel capacity (number of positions) writing is executed by a so-called half-channel sequence, which represents a channel sequence $\frac{1}{2}$ CS with all odd pulses excluded. Thus, instead of 128 channels with a channel capacity of $2^{14} - 1$, 64 channels are allocated to the additional track with a capacity of $2^{28} - 1$ readings each (about $2.68 \cdot 10^8$ pulses).

To simplify the registering circuits, these tracks do not have prestorage buffer blocks. In this case the write speed in each channel must not exceed 47-48 registered pulses per second (i.e., it must not exceed the main drum memory cycle).

The network for channel distribution of the monitor circuits consists of two blocks (see Fig. 1): relay block No. 1 for remote control of the identification number associated with a group of channels on the track, and distributor block No. 2. Control of relays R_1-R_3 of block No. 1 is executed from the panel of the physical instrumentation along the strands of a 50-pair cable (serving several laboratory user groups).

The states of the contact groups of the three relays are such as to permit eight different combinations of binary code in the form of 0/-2 voltages in the three output circuits of the relays for each user group. These circuits connect to five comparison circuits (block No. 2) allocated to each user group. A separate comparison circuit, or matrix, is similar to those used in the digital printer or in the buffer memory sections. A central-station code (without the lowest position 2^0, since only 64 channels are used) is sent to the left inputs of these circuits via six buses, and the values of the potentials from the relay contact circuits of block No. 1 are sent to the right inputs of the matrix. The lowest positions of the comparison circuits (2^2, 2^3, 2^4) for each of the five matrices are interlaced so that coincidence of the central-station code with the code for the group of five matrices for a given "value" of the potentials on the relay contact circuits yields coincidence for the group of adjacent channels of the track through every other one (e.g., 1-3-5-7-9).

Each comparison matrix can only have an output signal if a gate pulse from the half-channel sequence is transmitted from the turns circuit (K_4-K_8) of each matrix. A turn represents the trigger fired by the pulse to be registered in a diode-triode pair. The trigger is deactivated

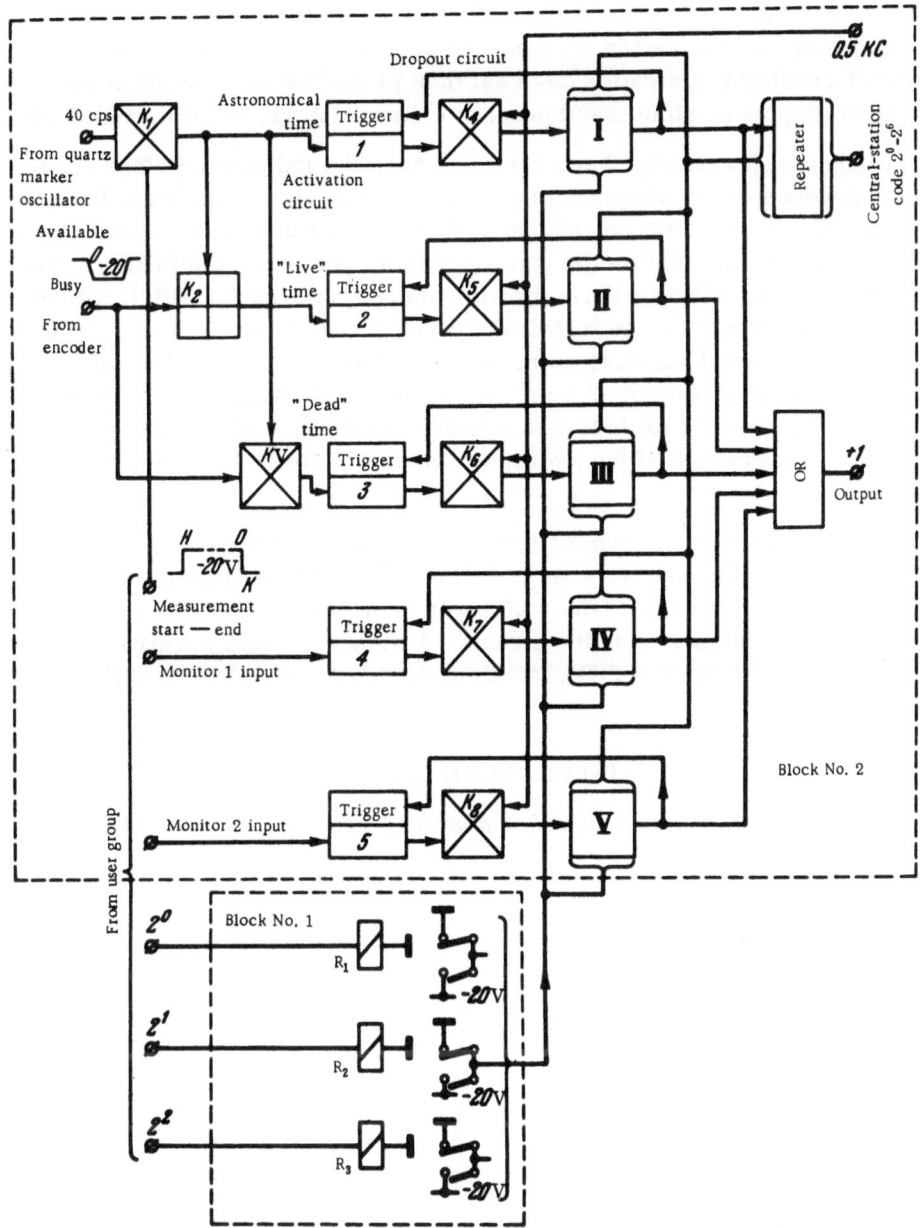

Fig. 1. Block diagram of the monitoring and time-measuring circuits.
I-V) Comparison matrices.

by the code comparison pulse in the matrix. A comparison event, i.e., activation of a blocking oscillator, comprises a WRITE "+1" command to the given channel of the monitor track.

For amplitude measurements the following variables are registered: a sequence of pulses from a quartz oscillator with a resultant frequency of about 40 cps, which is turned on at the beginning of measurements by gate K_1, which is controlled from the instrument panel of the appropriate user group (astronomical time). The same sequence, transmitted through the two gates K_2 and K_3, which are controlled by the state of the encoding device, permits registration of the "live" and "dead" times of that device according to its unoccupied and busy states (gates K_2 and K_3 operate). The cited frequency of the reference pulses, 40 cps, is adequate for measurement of these times with high statistical accuracy.

With changeover of the group code (from the circuits of the relay contacts) all five readings of the monitors are registered in the other five channels of the track. In this way one track replaces 64 direct-reading registration devices, thus providing data output in any mode available in the MRC system (digital printout, card punch, or direct interface with off-line computer).

For amplitude spectra having characteristic peaks the indicated method of measuring the "live" time can introduce error when such spectra are registered with large loads accompanied by heavy counting errors. As a matter of fact, with monotypical codes present at the encoder output the buffer memory will contain the codes for these peaks predominantly. Inasmuch as the output of these codes for writing in the main drum memory occurs in a definite phase of the latter, the encoding device will frequently be inhibited during the period of the main memory cycle. In the final analysis the period of the quartz oscillator and inhibiting period of the encoder will not be statistical independent, and their coincidence time will be characterized by the period of the resulting beats. In order to exclude this effect we have devised a random time-interval generator with strict observance of the average number of pulses; this device is discussed in another article [1].

To diminish the losses in the communication lines all monitor and illumination circuits are connected by coaxial telephone lines in a 50-line cable. Cable transfer is realized through capacitors to matching transformers (150 × 50 turns on an annulus 10 mm in diameter, with $\mu = 1000$) with a diode-resistance damping circuit. A standard cable permits the transmission of pulses with a leading rise time of 2-3 μsec along adjacent coaxial lines without perceptible noise.

LITERATURE CITED

1.   A. N. Volkov, A. E. Samsonov, and I. V. Shtranikh, Trudy FIAN, 42:90 (1968). [This volume, p. 92.]

CIRCUITS FOR THE REMOTE SELECTIVE DISPLAY AND "CLEARING" OF MEMORY "TRACKS"

A. M. Klabukov, A. E. Samsonov, and I. V. Shtranikh

The system for remote display and "clearing," i.e., the removal of stored data from each track, is based on diode selection matrices. The instrument panel for the user group has an additional five coaxial telephone lines for this purpose (in a common cable with the other groups). Two lines (four wires) permit selection of sixteen tracks in binary code. For this the potential of the cable wires (0 or -20 V) is inverted again at the input to the selection matrix, and then in

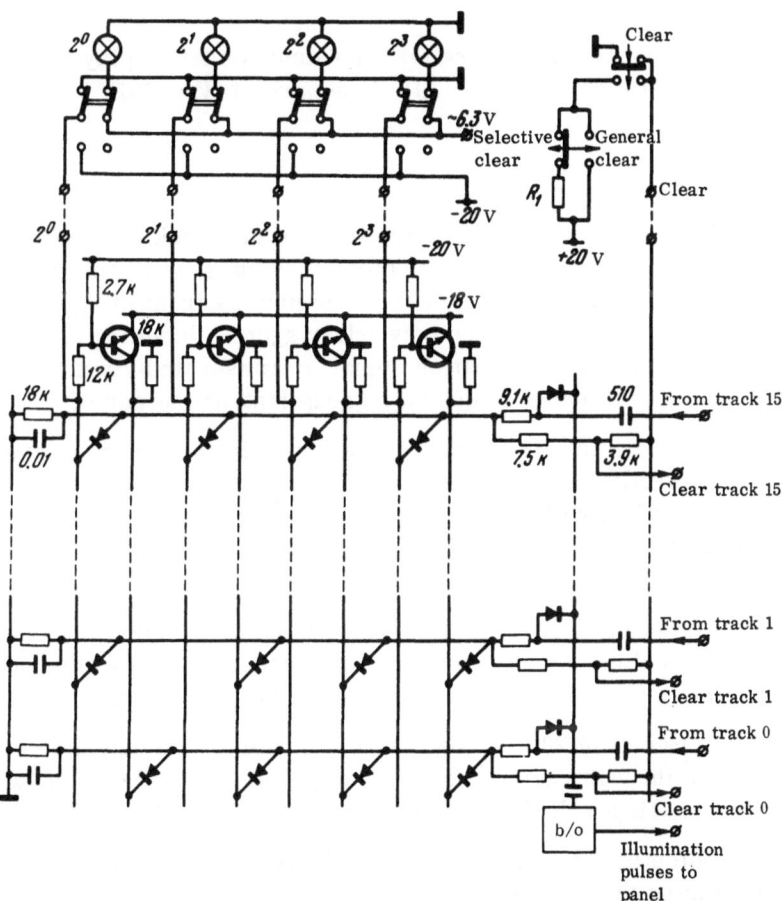

Fig. 1. Network for remote selective display of the "track" contents and "clearing" of memory.

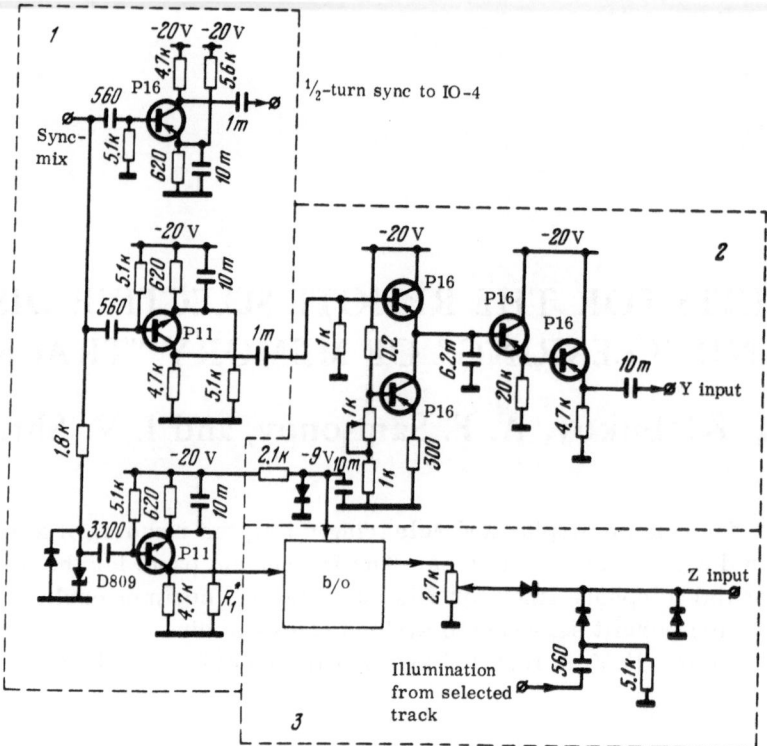

Fig. 2. Accessory network for one-dimensional display of the "track" contents. All diodes are of type D9E.

an ordinary diode network linear sampling of one input of the track logic block is made. The matrix output which does not carry a negative potential is chosen, as only it controls the given diode-resistance network for transmission of the code signals from the given logic block (Fig. 1). The common output of the transmission network triggers a blocking oscillator, and from the output winding of the latter an illumination signal is transmitted via a third coaxial line to the display network, which is situated in the user group instrument panel.

In addition to remote display, from each track logic block through a diode there is a circuit with trigger arm controlling the write gate-pulse transmission network. If the potential of this circuit is made positive, with an amplitude on the order of 0.5-2 V, the trigger of the logic block closes the transmission network [1], preventing regeneration, i.e., erasing the record.

For implementation of the foregoing the linear output of the matrix, besides controlling the illumination signal selection network, provides for transmission of the energizing potential to the diode circuit in the arm of the selected trigger of the logic block.

When a positive potential of 20 V is sent from the user group instrument panel via a seventh wire of the cable through resistance R_1 (Fig. 1) only the diode of the selected trigger is open, the remaining 15 trigger diodes for the other tracks remaining closed.

When a positive potential is transmitted along this wire of the cable without resistance, all diodes are open, causing the contents of all tracks of the given memory group to be general-cleared to zero.

The Display Block

In order to reduce the bulk of equipment on the user group control panels, as explained above, one-dimensional spectra are read out.

The display system comprises a standard IO-4 oscilloscope and a compact associated network (Fig. 2). The latter includes a linear sweep generator 2 and illumination network. For control of the given block a set of sync-mix pulses, including the half-turn, channel, and arithmetic pulse sequences, is transmitted along one cable from the MRC. In the dividing network 1 are separated the half-turn pulses for triggering the driven sweep of the oscilloscope, the channel pulses for triggering the line generator 2, and the arithmetic sequence for triggering the blocking oscillator 3 for illumination of the digital net. The code pulses are sent along another cable connected to the secondary winding of an intermediate transformer housed with the user group and connected to the telephone line from the track selection matrix.

LITERATURE CITED

1. A. M. Klabukov and I. V. Shtranikh, Trudy FIAN, 42:53 (1968). [This volume p. 53.]

AMPLITUDE SPECTROMETRIC SUBSYSTEM OF THE MEASUREMENT AND REGISTRATION CENTER

I. V. Shtranikh, A. M. Klabukov, and A. E. Samsonov

The problems involved in undistorted or linear transmission of signals whose configuration is a registration parameter play a vital part in the organization of centralized measurements. In particular, for amplitude measurements under such conditions the linear transmission of signals, absence of transmission noise, and the elimination of reflections in the cable transfer of signals are all significant factors.

The subsystem developed to meet these requirements in the Atomic Nucleus Laboratory of FIAN includes a cable line connecting the output devices of the detector circuit and the MRC input and terminating in special transformers.

Type RK-1 braided cables 100 m in length are connected in the instrument panel groups to the secondary winding of a special step-down transformer (see below). Passing through the MRC distributing frame, which consists of a cable-plug patchboard without common connection between the shielding of different cables, these circuits terminate in analogous input step-up transformers mounted in the amplitude encoders. A bipolar signal-shaping system is used for cable transmission. This system eliminates the constant component in transmission, a feature that is important under conditions of precision amplitude registration and especially for operation with statistically time-distributed pulses; the system also excludes long transient processes after each pulse and permits the use of transformer circuit coupling with low transformer inductances.

The following capabilities are afforded in conjunction with the transformers:

1) proper impedance matching of the output stages operating into the cable with the input stages at the receiving end of the cable;

2) absence of galvanic coupling between the user panels and MRC (otherwise equalizing currents at the line frequency might be created along the cable shields, resulting in additional interference and distortion of correct signal discrimination);

3) intelligent use in some cases of bias voltages and other connections in the presence of an insulated transformer secondary in the encoder circuits;

4) easy change of polarity of the working signals;

5) with the bipolar voltage form, whose first half-wave is chosen as the passive one, sharper alignment of the shaping device used to determine the zero-time reference in the encoder, which is necessary for inhibition of the input circuits and the start of signal-to-code conversion.

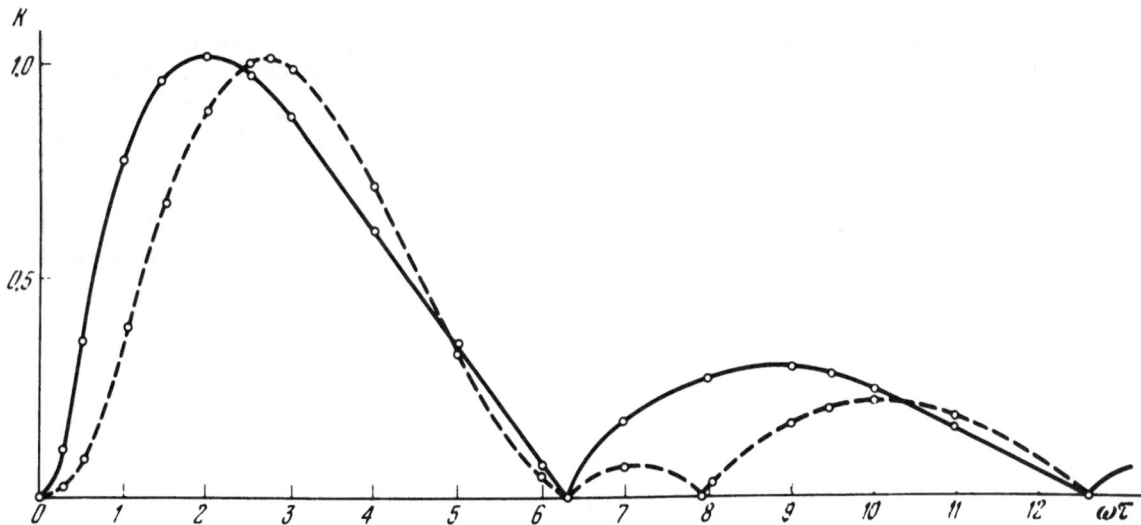

Fig. 1. Frequency characteristic of a component consisting of a differentiating and an integrating RC-circuit with equal time constants τ and one shorted delay line, $\tau_{\Lambda_1} = 0.5\tau$ (solid curve); the same but with the addition of second delay line, $\tau_{\Lambda_2} = 0.4\tau$ (bipolar shaping, broken curve). K) Relative transmission coefficient.

The disadvantage of bipolar shaping is a twofold increase in the signal duration and an increase in the noise factor of the preamplifier, detector, etc., due to superposition of part of the noise spectrum which is incompletely correlated in time. The first factor can increase the number of spectrum-distorting superpositions in the event of large loading. The second factor, given a proper choice of shaping parameters, still increases the signal-to-noise ratio by 15-20%.

It is well known [1] that the minimum noise level is attained for equal time constants τ of the differentiating and integrating circuits. If in addition single shaping on a shorted delay line is used, then for twice the signal transit time along the line, equal to the signal rise time or, in this case, τ, the signal amplitude does not decay. Owing to the narrowing of the pass band, the noise factor is diminished by a factor of $\sim 1/\sqrt{2}$. With further double shaping of the signal with the same delay the pass band will be narrowed somewhat more, as $\sin^2 \omega\tau$, instead of $\sin \omega\tau$, but due to the more pronounced property of the entire system of holding part of the preceding noise component until the occurrence of a signal, the signal-to-noise ratio becomes worse. In the limit as the duration of one of the shaping stages becomes very large, these components will not exhibit autocorrelation, and the signal-to-noise ratio will again suffer a $1/\sqrt{2}$-reduction.

Applying a second shaping chosen so as to suppress the second and succeeding maxima on the frequency characteristic of all these components (Fig. 1), i.e., using a second shaping with resonance at $\sim 1/4 \lambda$ in the vicinity of the frequency of the second maximum of the first shaping (shorter line), one would expect the signal-to-noise ratio to be improved. In this case, however, there is an additional simultaneous drop in signal amplitude and in the net result the signal-to-noise ratio remains unchanged.

Another article will be devoted to the problem of choosing the optimum values of the shaping parameters.

For the matching transformers we use an assembly consisting of the frame from a type SB-3 core, on which we have six simultaneously wound wires, with different numbers of turns in each of three sections of the frame (e.g., 15, 12, and 8 turns). Three wires of the windings are connected in parallel, the other three in series. The end result is a system of windings with a coupling coefficient greater than 0.95 and transformation factor 1:3. The frame and windings are housed in the core (without plug) and through the opening for the plug are completely wrapped with type KhVP 0.1 ribbon. With this construction the pulse front is transmitted due to the large direct mutual coupling of the windings. The high inductance of the windings is provided by the presence of a carbonyl core, which is lossless for the high-frequency components of the pulse front.

The additional core and KhVP ribbon winding greatly enhance the inductance of the transformer windings, but without distorting the pulse front, because, clearly, the magnetic flux controlling transmission of the front is shielded by the SB core. Without the SB core the front of the transmitted pulse suffers appreciable deterioration.

Standard networks are used for shaping of the bipolar pulse, including double shaping with shorted lines. The first shaping occurs at the preamplifier output, the second at the input to the cathode follower operating into the transformer (through capacitive coupling). The amplitude losses in the output signal are eliminated in the second version by means of a balanced stage with a delayed signal in one leg.

Using the low-noise preamplifier network described above in conjunction with integration and differentiation time constants of 5 μsec each in the main amplifier and a shaping period of 2.5 μsec in the lines, we obtain a noise factor at the 3-keV level.

LITERATURE CITED

1.   A. B. Gillespie, Signal, Noise, and Resolution in Nuclear Counter Amplifiers, Pergamon, London (1953).

DIFFERENTIAL AND INTEGRAL COUNTING BLOCK
FOR SPECTRUM STABILIZERS

A. N. Volkov and I. V. Shtranikh

The operation of most present–day digital spectrum stabilizers [1–3] is based on stabilization of the position of a reference peak in a particular group of channels. The reference peak
is divided into two halves by means of special digital "windows." The difference between the
readings in the "windows," referred to their sum or square root of their sum, characterizes
the degree of shifting of the reference peak from predetermined limits. Our proposed network
(Fig. 1) is designed for digital spectrum stabilizers and permits the difference and sum of the
readings in the "windows" to be determined either in analog or digital form. The network

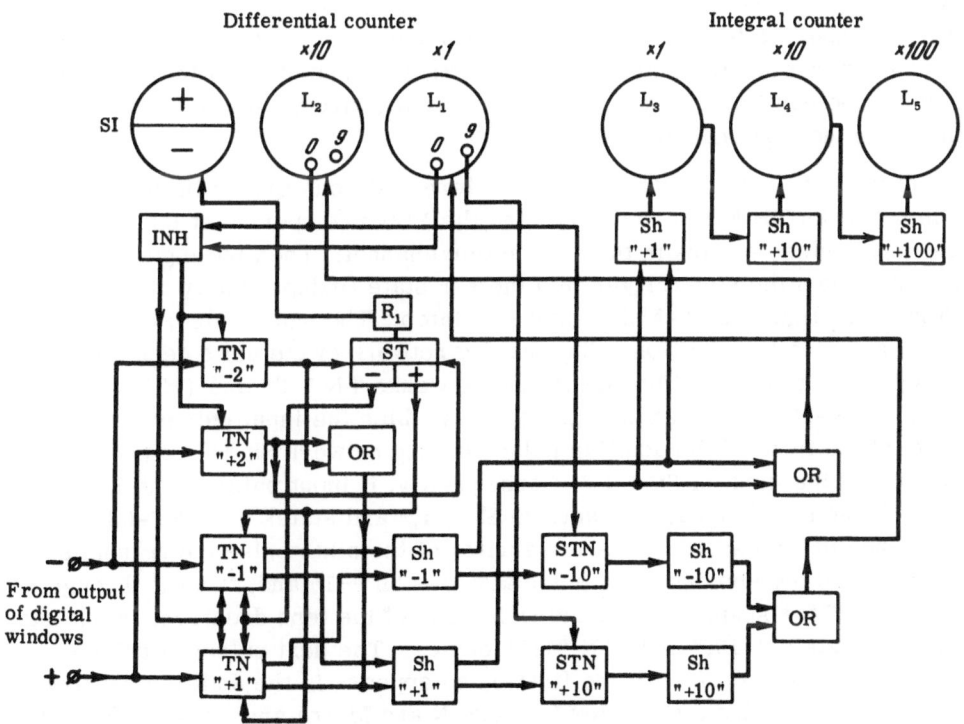

Fig. 1. Block diagram of the differential and integral counting block. Sh "±1,"
Sh "±10," and Sh "+100" are identical networks for shaping of the start pulses
to the decatrons for entry of the numbers ±1, ±10, and +100 into the counters.

89

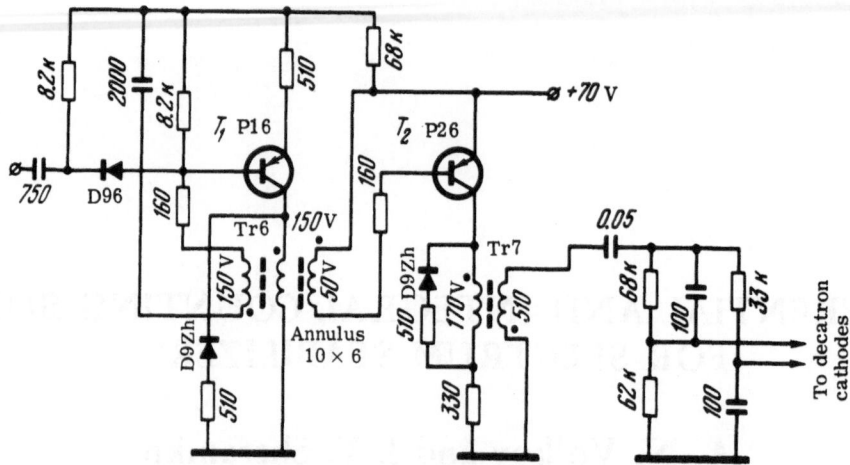

Fig. 2. The shaping network.

contains a decade three-position integral pulse counter (L_3, L_4, L_5) and a two-position differential counter (L_1, L_2).

Both counters are made with type A103 switching decatrons operating at 50 kc and have a transistorized drive. The pulses from the left and right digital windows are transmitted to the corresponding addition or subtraction channel transmission network (TN "+1", TN "-1"). The output signals TN "+1" and TN "-1" are fed into the networks (T_1, T_2) for shaping of the addition or subtraction pulses (Sh "+1," Sh "-1") for actuation of the units decatron (L_1) in accordance with the states of the decatrons (L_1, L_2) and the sign trigger.

The sign trigger is controlled by the input pulses of the counter through the transmission networks TN "+2" and TN "-2," which open only for the zero position of both decatrons of the differential counter through the inhibiting network (INH). The state of the sign trigger (ST) uniquely governs the position of the contacts in relay R_1, through which the indices "+" or "-" are illuminated on the sign indicator (SI).

In the initial state both decatrons of the differential counter are in the "0" position, the transmission networks TN "±1" are closed, and the "±2" are open. Let us say for definiteness that the first input pulses occurs in the subtraction channel. Then the input pulse, on passing through TN "-2," takes the sign trigger into the "-" state (independently of the preceding state) and through relay R_1 lights the "-" sign indicator lamp. The same pulse, on passing through the OR circuit, actuates the addition pulse shaping network of the units decatron (Sh "+1"), and the decatrons go over to the "-01" position, the networks TN "+2" and TN "-2" are closed, and in TN "+1" and TN "-1" those branches are open by which the input pulses of the subtraction channel arrive at the input of the addition pulse shaping network Sh "+1" and the addition channel pulses arrive at the input of Sh "-1." Consequently, if input pulses continue to be transmitted in the subtraction channel, they actuate Sh "+1," and states "-02," "-03," etc., are recorded in sequence in the decatrons. After arrival of the ninth pulse the transmission network STN "+10" opens, and the ensuing tenth pulse actuates Sh "+1" and, from it through STN "+10," the network Sh "+10" for shaping of the addition pulse of the tens decatron. From state "-09" the decatrons go to state "-10," and STN "+10" closes. The next pulses fill the ones and tens decatrons in the same sequence until the first pulse appears in the addition channel. Suppose that the decatrons were in the state "-NM" (where N and M are any integers from 0 to 9). Then the input pulse of the addition channel is transmitted through the open TN "+1" to Sh "-1," where a units decatron subtraction pulse is shaped. The decatrons go over to the state "-N(M - 1)." After arrival of M pulses in the addition channel the decatrons are found in the state "-NO," TN "-10" opens, and the (M + 1)th pulse is transmitted to the network for shaping of the tens decatron subtraction pulses; the decatrons go over to the new state "-(N - 1)9." If at this time another 10(N - 1) + 9 pulses are sent into the addition channel, the state of the decatrons will be "-00," TN "-2" and TN "+2" reopen, and TN "+1" and TN "-1" reclose.

The next addition channel pulse takes the sign trigger and sign indicator into the "+" state, actuates Sh "+1," and causes the decatrons to go into the "+01" position, etc. The integral counter consists of three series-connected decatrons (L_3, L_4, L_5) operating only in the addition mode.

The entire logic and actuation network is transistorized and has a simple and practical indication and control system. The network is mounted in a unit with dimensions 180×450 mm. The front panel of the unit consists of two rows of decatrons and two switches. The upper row forms the indicator of the differential counter and comprises two decatrons (L_1, L_2) and an unbalance-sign indicator. The lower row of decatrons (L_3, L_4, L_5) is the state indicator of the integral counter. The differential counter switch enables one to set the limits of admissible unbalance in the readings of the left and right peak windows with a certain sum of the readings as determined by the switch of the integral counter. If the unbalance exceeds the limit settings, a start signal for the reference peak position correction system appears at the network output. A similar signal proportional to the degree of unbalance is taken from the ninth cathode of decatron L_2, and a signal proportional to the sum of the readings in the two windows is taken from the ninth cathode of decatron L_5. The coded form of the unbalance signals is taken directly from the decatron cathodes, and the sign of the unbalance is determined by the state of the sign trigger.

LITERATURE CITED

1.   I. A. Ladd and J. M. Kennedy, Clark River Report, Ontario, CREL, p. 1963 (1961).
2.   R. A. Dudley and R. Scarpatetti, Nucl. Instr. Meth., 25:297 (1964).
3.   M. Nakamura and R. L. Lapierre, Nucl. Instr. Meth., 32:277 (1965).

GENERATOR OF RANDOM NUMBERS, TIME INTERVALS, AND PULSE AMPLITUDES

A. N. Volkov, A. E. Samsonov, and I. V. Shtranikh

The adjustment and testing of modern multichannel amplitude analyzers and time-base pulse selectors for nuclear physics require the availability of a pulse generator with an equal-probability distribution of pulse amplitudes and time intervals between pulses (within prescribed amplitude and interval limits).

Existing generators normally have a linear variation of pulse amplitude at constant time spacing. The generator described below is the culmination of an attempt to build a more universal instrument for the testing and adjustment of various electronic devices.

The generator is capable of operating in one of the following modes.

1) precise-amplitude pulse generator; the amplitude is set by means of a ten-position switch from 0 to 10 V, correct to ±0.5 mV;

2) generator of pulses with linearly-increasing amplitude (from 0 to 10 V);

3) generator with equal-probability distribution of pulse amplitudes in the range from 0 to 10 V.

In each mode it is possible to set one of three signs for the distribution of time intervals between pulses: a) constant intervals between successive pulses; b) linear growth of successive intervals between pulses in the range from 0 to T; c) equal-probability distribution of intervals between pulses in the range from 0 to 2T, where 1/T is the mean repetition rate of the pulses; this frequency is set on a quartz frequency divider between the limits 10 cps and 10 kc.

The generator provides for the realization of a 10-position coded pulse form. This permits the simultaneous introduction of code and analog pulse forms into the analyzing system, a feature that is very useful for adjustment of the equipment.

The random number generator (see Fig. 1) contains the following functional components:

1) a source of random pulses with a Poisson time distribution, including a radioactive emission source (R), nuclear radiation detector (NRD), amplifier, and phasing network (Ph);

2) a shift register comprising a circuit of series-connected tunnel diode triggers (Tr1-Tr10);

3) a buffer memory register (BMR) using scaling sections (SS1-SS10), where input and storage of the random code is executed;

4) transmission networks (TN1-TN10) for switching of the timing pulse sequence to the inputs of the scaling sections;

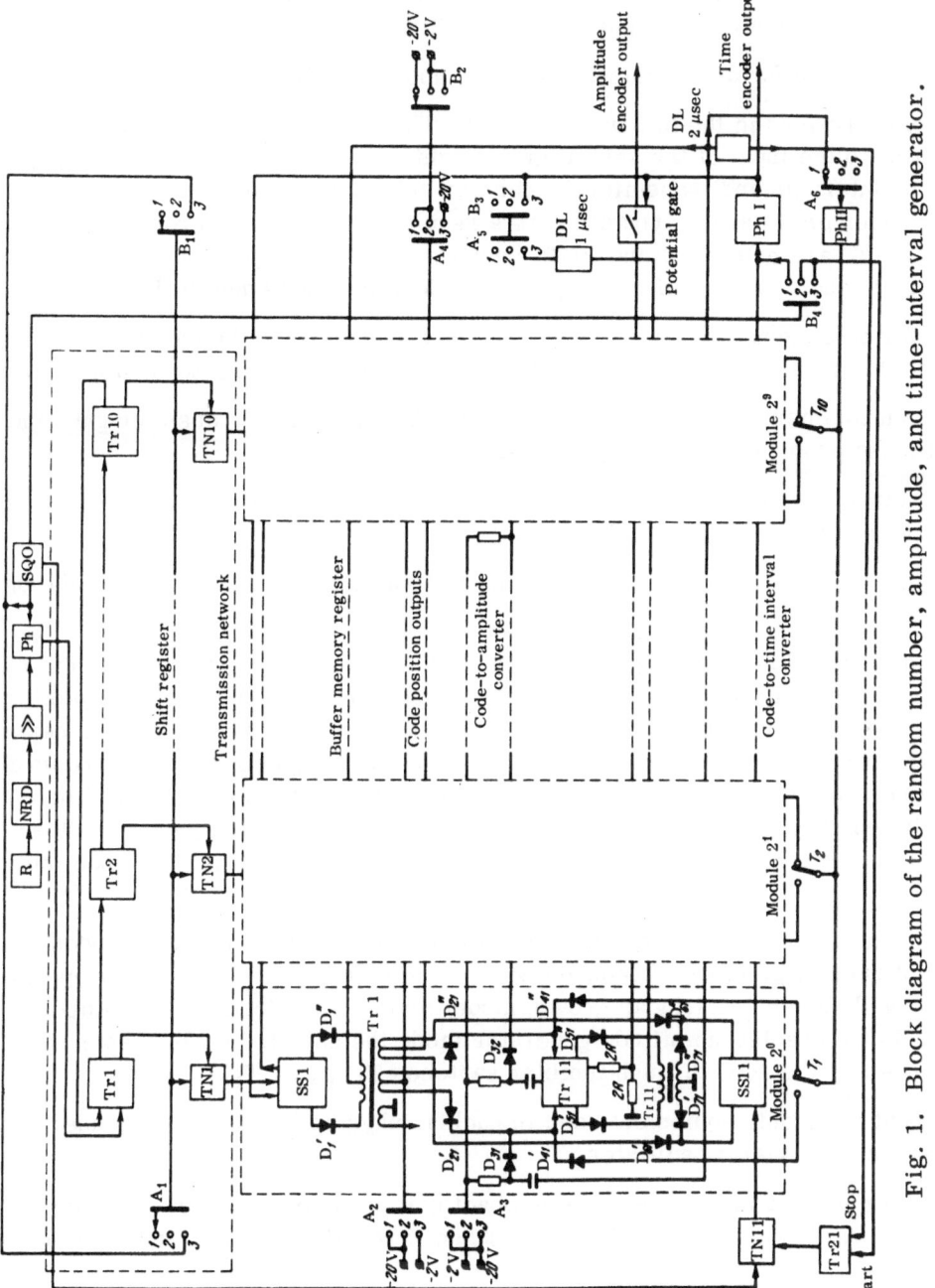

Fig. 1. Block diagram of the random number, amplitude, and time-interval generator.

5) a code-to-analog converter (CAC) executing transformation of the random code into the appropriate pulse amplitude;

6) a code-to-time interval converter (CTC);

7) a standard quartz oscillator (SQO) with frequency divider permitting variation of the pulse rate from 10 cps to 10 kc; the oscillator generates a sequence of phasing pulses for input to the BMR and a sequence of carrier pulses for the CTC register;

8) switches A and B for activation of any of the above-mentioned generator modes.

The network functions in the following manner. The pulses from the random signal source [1] are transmitted to the input of the shift register, which opens gates $TN_1, \ldots, TN_{10}$ in sequence over random time intervals and transmits the carrier pulses of the phasing sequence from the SQO to the buffer memory position triggers. As a result, random equal-probability 10-position numbers from 0 to 2^{10} are entered in the BMR, because in this case the probabilities of the input of zero or one in each position of the register are identical and equal to $\frac{1}{2}$.

The transfer pulse generated by the CTC register at the end of the carrier sequence translates (mode 3) the random code of the BMR into the amplitude code-to-analog converter.

The CAC, on command from the shaping network (Sh1), opens a transistor gate and generates an output pulse proportional to the recorded code.

The next start pulse received from the SQO begins a new conversion cycle (the frequency of the start pulses can be varied from 10 cps to 10 kc).

The time interval from start to the appearance of an output pulse depends on the operating mode of the CTC.

For a random time interval [2] (equal-probability distribution of intervals in the range from 0 to T) the random number code is first transferred from the CAC register into the CTC register. Then a new random number code is admitted into the CAC register. The next start pulse initiates the random time-interval generation cycle, at the end of which a random amplitude pulse, uncorrelated with the random interval code, is generated in the CAC. The correlation existing between the present interval and amplitude of the preceding pulses is close to zero due to the introduction of cross-connections in the translation of the CAC code into the CTC.

In the constant or linearly-increasing pulse amplitude (or time-interval) modes of operation the source of random pulses is disconnected, and the buffer memory register operates as an accumulative scaling network (linear growth mode) or is cut off (constant amplitude and time-interval mode). In the precise-amplitude generator mode the selected code is entered by each start pulse at once into the PAC through tumbler switches $T_1, \ldots, T_{10}$.

The standard electronic equipment used in nuclear physics laboratories is used for the random pulse source and quartz oscillator.

The generator is built in the form of a compact unit of dimensions 90×450 mm, consisting of 12 modules (position modules $2^0, 2^1, \ldots, 2^9$, shift register module, and logic network module).

LITERATURE CITED

1. E. I. Dolgirev, P. M. Maleev, and V. V. Sidorenko, Nuclear Radiation Detectors, Sudprom-giz, Moscow (1961).
2. I. V. Dunin-Barkovskii and N. V. Smirnov, Probability Theory and Mathematical Statistics in Engineering, Gosénergoizdat, Moscow (1955).

CALORIMETRIC DEVICE FOR MEASURING
THE POWER OF SEMICONDUCTOR LASERS

B. D. Kopylovskii, V. S. Bagaev,
V. S. Ivanov, and A. V. Slavov

In studying the physical properties of semiconductor lasers it is required to measure the luminosities in the spontaneous emission regime, as well as the generation regime.

The registration of luminous energy is based on its conversion into a form of energy that is directly measurable. Thermal radiation detectors stand unexcelled over other sensors by virtue of their nonselectivity over a wide spectral range. A representative example of these detectors is the "black body" type of calorimeter [1].

The calorimeter was constructed in the form of an evacuated glass cylinder with a small entrance window in the end, made of sapphire. Two cones with identical parameters were placed in the cylinder, one as the working body for radiation detection, the other as a compensation body. The sensing elements used to register the heating of the cone were type T8S1M thermistors. The output leads from the thermistors were platinum wires 0.01 mm in diameter. The thermistors were attached right to the cone by means of heat-conducting resin. The copper cone was fabricated by an electrolytic method. This method was used to produce cones with minimal mass, ~10 mg. For better radiation absorption the cones were blackened on the inside

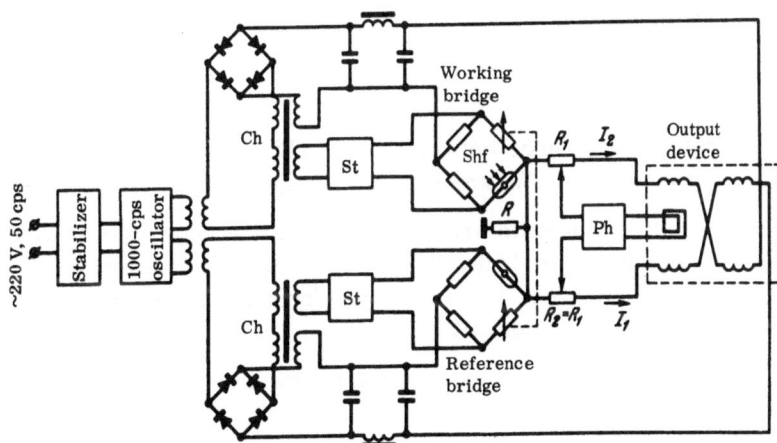

Fig. 1. Block diagram of the thermistor bridge. Ch) Saturation choke; St) reference- and working-current stabilizers; Ph) measurement photocompensator.

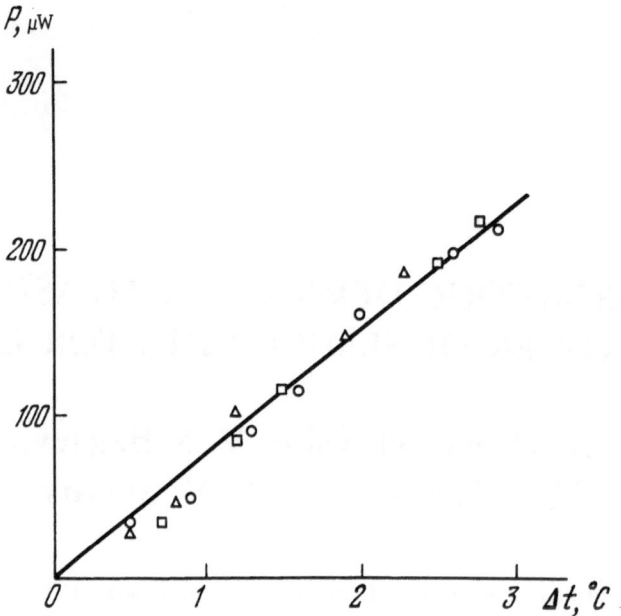

Fig. 2. Calibration curve for the determination of α.

with polysulfide compounds. The calorimeter does not require additional calibration and can be used as a monitoring device and has the important virtue of being able to register energy from hundredths of a joule.

The energy absorbed by the cone is determined by the expression

$$\Delta E = cm\Delta t \,[\,\text{cal}\,], \tag{1}$$

where c = 0.09 cal/g·deg is the specific heat of the copper cone and m is the mass of the cone.

The heating of the cone Δt was determined by a bridge technique. For the measurements a type MTO-1 thermistor bridge was used; it is designed to determine low shf powers. The device indicates the dc power equivalent (in terms of effect on the thermistor) of the measured high-frequency power. A block diagram of the bridge is shown in Fig. 1.

For low heating of the cone the following relation is valid:

$$\Delta t = \alpha P, \tag{2}$$

where Δt is the temperature difference between the working and compensating cones, P is the reading of the bridge instrument in μW, and α is the temperature coefficient.

The values of α for three thermistors were determined experimentally by means of an ultrathermostat and the thermistor bridge. The results of the measurements are presented in Fig. 2. The temperature coefficient α is determined from the slope of the line and for the given of thermistors turns out to be 0.013 deg/μW.

Substituting Eq. (2) into (1) and allowing for transmission, absorption, and reflection from the entrance window, we write

$$\Delta E = 4.18\gamma cm\alpha P, \tag{3}$$

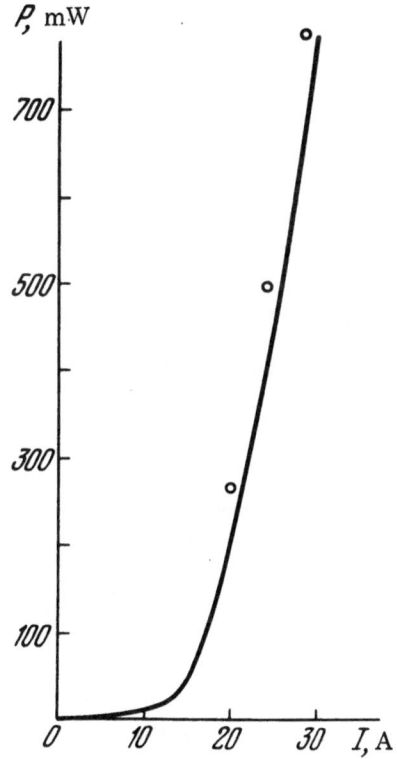

Fig. 3. Radiant power output of a GaAs laser versus the current measured by the photoelectric method. The dots indicate the power values measured with the calorimeter.

where ΔE is the measured energy in joules, α is a coefficient accounting for reflection, absorption, and transmission at the window (α = 0.013 deg/μW), and P is the reading of the instrument in μW. Equation (3) does not take account of the thermal losses due to attachment of the cone, the thermistor leads, or heat conduction of the rarefied air in the calorimeter (pressure = 10^{-5} mm Hg).

It is apparent from Eq. (3) that the sensitivity of the calorimeter is determined by the mass of the cone and sensitivity of the thermistor bridge. The minimum reading that can be reliably made with the bridge is equal to 1 μW. The mass of the copper cone used in the calorimeter is 50 mg, hence the minimum detectable energy is $1.7 \cdot 10^{-4}$ J.

Inserting all known data into Eq. (3), we find

$$E = 1.71 \cdot 10^{-4} \; \mu w. \tag{4}$$

Expression (4) was used to compute the radiant power.

Some comparative measurements of the radiant power output of a GaAs laser by the photoelectric method and with the calorimeter demonstrated satisfactory agreement of the results (Fig. 3). Moreover, measurements were conducted to ascertain the power output of an InSb laser (λ = 5.3 μ), which turned out to be 250 μW.

In conclusion we would like to thank E. V. Gorskin, Yu. A. Korolev, and L. M. Novak for assisting with the construction of the calorimeter.

LITERATURE CITED

1. R. A. Kazaryan, É. S. Vardanyan, and F. R. Saforyan, Pribory i Tekh. Éksperim., No. 3, p. 166 (1964).

INFRARED TELEVISION MICROSCOPE FOR THE INVESTIGATION OF RECOMBINATION RADIATION AND THE OPTICAL PROPERTIES OF SEMICONDUCTORS

N. L. Artem'ev, V. S. Bagaev, O. V. Gogolin, Yu. A. Efimov, and B. D. Kopylovskii

In the present article we describe an infrared television microscope suitable for investigations of the recombination radiation and optical properties of semiconductors in the wavelength range from 0.5 to 2.3 μ.

Television is enjoying growing applications in scientific research [1, 2]. Video techniques make it possible to observe low-contrast objects with large magnification. The television system permits amplification and regulation of the contrast and brightness by purely electronic techniques, and the use of a transmitting tube sensitive in the infrared domain enables one to study the optical characteristics of a number of semiconductors.

Fig. 1. Exterior view of the infrared television microscope.

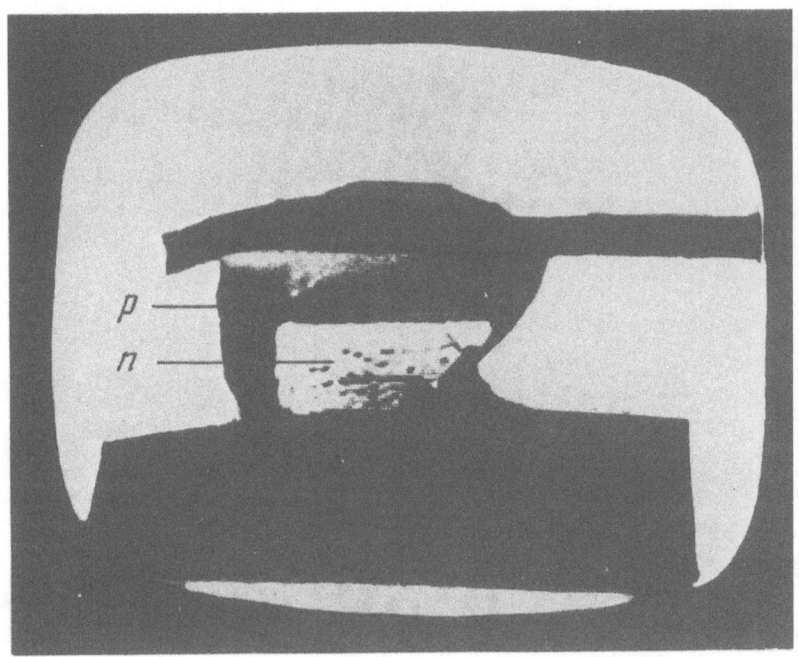

Fig. 2. Transillumination image of a GaAs diode in infrared radiation.

The researcher may often be concerned with special parts of the image. Video techniques can be used to enhance the resolution of individual details of the object under study. This can be done by deliberate "distortion" of the transmitted distribution of luminance gradations in order to highlight details of the image. This problem can be solved by regulation of the amplitude characteristic of the video system.

The infrared television microscope consists of an industrial television apparatus of the PTU-4A type with a transmitting camera modified for an infrared vidicon, and a Karl Zeiss type "Nu" universal polarization microscope. An overall view of the device is shown in Fig. 1. It uses a type LI-405 infrared vidicon, which is sensitive in the wavelength range from 0.5 to 2.3 μ, as the photosensitive element. This kind of vidicon comprises a television transmitting tube with a photoresistance whose target is made of sulfur-treated lead monoxide.

The power threshold for the LI-405 tube at an absolute black body temperature T = 300°C is at worst 10^{-3} W/cm^{-2} for a signal-to-noise ratio of one. The shape of the spectral characteristic of the vidicon can fluctuate, depending on the technique of fabrication of the tube, but its characteristic points, i.e., the maximum and "red" and "blue" limits, stay right around 1, 2.3, and 0.54 μ, respectively.

A comparison of the resolving powers of an electron-optical image converter (EOC) and the vidicon under equal conditions (i.e., with normal illuminance of the screen and a line contrast close to unity) yields the following data:

Resolving power at image center: 30 lines/mm for the EOC; 30 lines/mm for the vidicon.

Resolving power at the edge of the image: 3 lines/mm for the EOC; 25 lines/mm for the vidicon.

The foregoing comparison indicates that the vidicon has a significantly higher uniformity of resolution over the image field. The main characteristic governing the applicability of

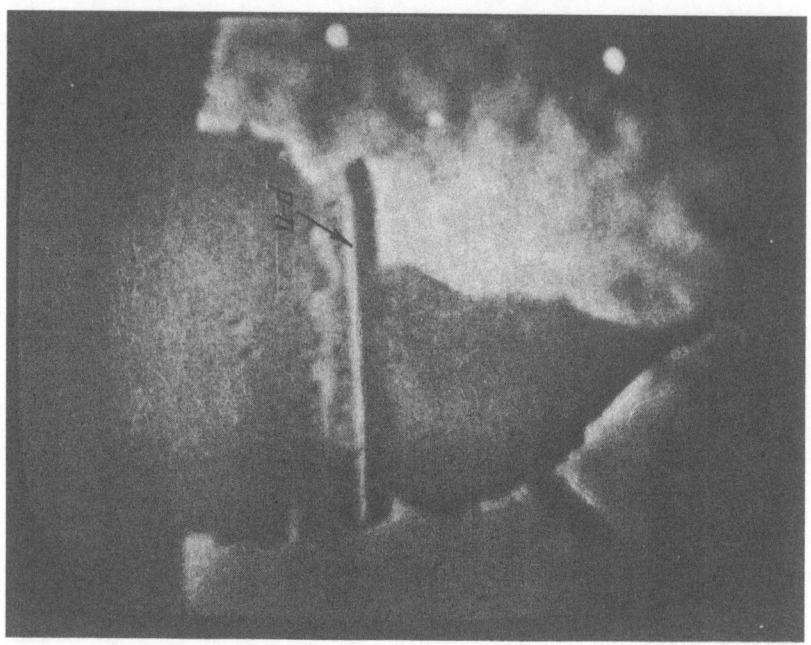

Fig. 3. Luminescence of a GaAs diode in the subthreshold regime.

vidicons in instruments designed for investigation of the physical properties of semiconductor materials is their high sensitivity in the near-infrared portion of the spectrum.

We used the television microscope described above to observe the luminescence of GaAs diodes both in the spontaneous emission and in the generation regime. A transillumination image of the diode in infrared radiation is shown in Fig. 2 at a magnification of 2000. The nontransmissive p-region and infrared-transmissive n-region, separated by the p-n junction, are well-resolved in the photograph. Surface defects on the cavity mirror are also clearly discerned.

For operation in the subthreshold regime the peak of the spontaneous emission spectrum for a GaAs laser corresponds to 9000 Å. The luminescence represents a narrow, uniformly luminescing band in the p-region of the diode in the immediate vicinity of the p-n junction, as indicated in Fig. 3.

For operation in the laser regime the radiation wavelength corresponds to 8440 Å (T = 71°K). For the observations we used adsorption filters with suitable characteristics to attenuate the powerful radiation impingent on the transmitting tube. On going over to the generation regime, bright spots appear against the background of the narrow luminescent part of the p-n junction, defining the zones in which conditions for generation are met.

We used special infrared polaroids to study the polarization of the GaAs laser radiation. The onset of radiation polarization in transition to the generation regime is clearly visible on the screen of the television microscope. The polarization vector **E** for different lasers can coincide with the plane of the p-n junction or it can be perpendicular to it. In some cases the individual generation spots can have mutually perpendicular polarization for the same laser.

The infrared vidicon used in our system has its maximum sensitivity in the spectral range from 1.2 to 2.3 μ. With the appropriate optics this permits the observation of recombination radiation from Si and other materials, a capability that is of tremendous importance for the further study of electroluminescence in semiconductors.

LITERATURE CITED

1. N. I. Murashov and G. D. Shnyrev, Photoelectric Infrared Polaroscopic and Flaw Detection Techniques in Semiconductor Materials, Collection edited by V. I. Grechushnikov, Izd. AN SSSR (1963), p. 40.
2. O. Deutschbein and M. Bernard, Solid State Physics, Semiconductors, 1:117 (1960).